HSE
Health & Safety
Executive

SOUND SOLUTIONS
Techniques to reduce noise at work

HSE BOOKS

HS(G)138

This guidance is issued by the Health and Safety Executive. Following the guidance is not compulsory and you are free to take other action. But if you do follow the guidance you will normally be doing enough to comply with the law. Health and safety inspectors seek to secure compliance with the law and may refer to this guidance as illustrating good practice.

CONTENTS

INTRODUCTION

Background

Despite rapid technological progress and changing employment patterns, hundreds of thousands of workers are exposed every day to loud noise at work. Many will suffer permanent hearing damage. The Noise at Work Regulations 1989 introduced legal requirements to reduce noise levels and so protect the hearing of workers. Surrounding publicity has raised awareness of health risks arising from exposure to noise at work.

A recent HSE survey found that most employers knew of their duties to control noise but many were failing to take action. One reason was a lack of understanding both about how to introduce effective measures and the techniques available.

Aim of the book

This book is for managers (although works engineers and safety representatives will also find it useful) and aims to show that noise problems can be solved in many ways. It offers real examples of how some companies have reduced noise at work. Although each industry has its own working practices, many noise problems and solutions are not unique and are relevant in several industries.

The case studies

The case studies in this book have been organised by source of noise. Each describes the nature of the noise problem, the solution applied by the company, the cost (at 1994 prices) and the noise reduction benefit achieved. Inevitably, some of the language is technical and so a glossary is included at the end of the book. The case studies are designed to give managers an idea of what is achievable and are not meant to reproduce technical manuals. The tables at the back of the book provide easy cross-reference to industry and to the method of noise control.

Check-list for managers

The book includes a check-list for managers to use when discussing alternative noise control methods. Many of the solutions described have been developed by employers' in-house expertise and a significant number were improved by involving employees and their representatives in discussing problems and trying out ideas. Some organisations found that employing noise consultants with wide experience in investigating excessive noise at work led to effective, value for money solutions. HSE has also published guidance on employing health and safety consultants.*

Acknowledgements

HSE commissioned Ian Sharland Ltd to gather the information for the case study material in this book. We would like to thank them and the companies and consultants who agreed to be involved with this publication (listed at the back of the book).

* *Selecting a health and safety consultancy* IND(G)133(L) HSE 1992

CHECK-LIST FOR MANAGERS

Use this check-list to help you decide which noise control technique might best solve a noise problem. Discuss the details with a noise control engineer.

1 Observe, listen to and touch (where safe) the machine in question

- ❏ What is the problem?
- ❏ Where is it?
- ❏ Is a vibrating part the source of noise?
- ❏ Am I sure about what is the main source of noise?
- ❏ Am I treating the most dominant noise source first? (There is little point in treating the easiest or cheapest first if it is not the main source of noise.)
- ❏ How many employees will benefit from the noise control measure?
- ❏ What will be the cost per protected employee?

2 Consider the source of noise

- ❏ Is it reasonably practicable to replace the machine by one with low noise emission?
- ❏ Could the machine be removed from the occupied working area without disrupting production?
- ❏ Could the working area be altered so that the main noise source is moved where it least affects employees?
- ❏ Could worn or faulty parts be replaced, particularly if the machine is getting increasingly noisier?
- ❏ Is it possible to modify parts of the machine, eg by replacing components with ones designed to operate more quietly?

3 Consider how the source radiates noise

- ❏ Is the source vibrating the machine's panels? Isolate the panels or add damping materials to them.
- ❏ Can vibrations be felt in the structure of the building? Isolate the machine from the building with isolation mounts or isolated foundations.
- ❏ Is the noise caused by continuous impacts from falling material? Add damping material to receiving trays and chutes etc.
- ❏ Are solid guards attached to the machine around noisy components? Line guards with absorbent material.
- ❏ Is the major noise source caused by either the inlet or exhaust of air or gas from the machine? Fit an appropriate silencer to the inlet or exhaust (or both).
- ❏ Is the noise caused by a sudden release of air from a compressed air system? Fit proprietary silencers or feed the exhaust away from the working area.

4   Consider the path of the noise

❑ Could you position the worker away from the noise source? (Doubling the distance can reduce the effect of noise by 6 dB.)

❑ Could you fit a suitably designed enclosure around a machine that does not require 'hands on' operation? Include access panels where periodic inspection is required.

❑ Could you acoustically treat openings in the machinery into which material is placed or from which the product is removed?

❑ Should you fit acoustic ducts or position quiet fans in enclosures where there may be a build up of heat?

❑ Could a noise haven be built for employees supervising the operation of large machines where enclosure is too difficult (eg a large printing press)?

❑ Could you erect barriers or screens between different elements in the production process to separate quieter operations from noisy ones? Add absorptive material to the building to soak up reverberant noise (echoes).

❑ Would active noise control be appropriate where the noise level is constant and is made up of a major low-frequency tonal component (for example, dryers or exhaust fans)?

5   Finally, measure the new noise levels after the control method has been fitted.

AVOIDING PITFALLS WHEN INTRODUCING NOISE CONTROL

The following techniques are described in the case studies. For further information, see HSE Noise guide No 4 *Engineering control of noise.*[*]

Damping
This involves adding material to reduce vibration.

❑ Is the damping material becoming detached from the structure or machinery?

❑ Is the damping material deteriorating?

❑ Remember the limited range of noise frequencies for which damping may be effective.

❑ Are you using enough damping material? The material should be at least as thick as the element or structure being damped.

Design
This is changing the total or partial design of the whole or a component of the machine.

❑ Consider the potentially high cost of redesign.

❑ Will the change in design of one component affect how others work?

❑ Will a change in one component's material result in the new material being incapable of taking the mechanical stress of the machine?

* *Noise at work: Noise assessment, information and control. Noise Guides 3 to 8* HS(G)56
HSE Books 1990 ISBN 0 11 885430 5

Screens and barriers

This involves placing an obstacle between the noise source and employees.

- ❏ They may be ineffective at low noise frequencies.
- ❏ They only reduce direct noise and will not stop reflective noise.
- ❏ Place the screen or barrier as close to the noise source or employee as possible.
- ❏ The screen or barrier should be made of a heavy material, eg steel or brick, and should be lined with absorptive material on the side facing the noise.

Enclosure

This involves placing a sound-proof cover over the noise source.

- ❏ Is the design of the enclosure correct? It should have:
  - ❏ good seals with no leaks;
  - ❏ inlets and outlets for material and services;
  - ❏ cooling ducts;
  - ❏ at least 50 mm of internal absorptive material.
- ❏ Can heat from the machine be extracted from the enclosure?
- ❏ Cover the absorptive material with a thin impervious film (eg plastic) to protect it where oil or water is likely to be in the enclosure.
- ❏ Fit the enclosure with self-closing devices on all access panels.

Isolation

This involves separating machinery noise from its surroundings.

- ❏ Ask for specialist advice from an anti-vibration mount supplier.
- ❏ The resonant frequency of the anti-vibration mount must not coincide with that of the machine.
- ❏ Make sure that the mounts are able to carry the weight of the machine.
- ❏ Watch for possible interference with any casing or enclosure caused by the machine's movement on its mounts.

Refuges

These are noise reduced enclosures for employees.

- ❏ Is there adequate ventilation?
- ❏ Are there good seals on the doors and windows?
- ❏ Are there self-closing devices on all doors?
- ❏ Is it constructed of a heavy material, with plenty of acoustically double-glazed windows? (Do not make it a hideaway.)
- ❏ Is it isolated from the floor?
- ❏ Is it large enough?

Active noise control

This is a computer-controlled noise reduction method (see glossary for an explanation). This can be:

- ❏ ineffective at medium to high noise frequencies;
- ❏ ineffective on widely varying noise sources;
- ❏ expensive.

Silencers

These are attachments fitted to the inlet or exhaust (or both) of a moving air or gas stream emitted from the machine.

- ❏ Absorptive silencers are usually only good for medium to high noise frequencies.
- ❏ Passive silencers are most effective in controlling the tonal frequencies they are tuned to reduce.
- ❏ Contamination by debris (dust or liquids) will reduce efficiency.

Room acoustics

Reduce reverberation by changing the acoustics of a room or work area with absorptive material.

- ❏ The potential cost may be high.
- ❏ Moisture in the atmosphere may cause deterioration.
- ❏ There may be a reduction in natural light where the absorbers are suspended from the roof.

Finally

Check that the noise level hasn't increased over time (ask the operator). This may mean that a control feature is not being used properly (eg an enclosure panel might not be in place) or may indicate a deterioration in the control materials.

TABLE OF CASE STUDIES

	Title	Noise source	Industry	Control method
1	Reducing noise in gravel chutes	Aggregate handling	Mining/quarrying	Damping
2	A lower noise alternative for compressed air drying	Air dryer	Automotive	Design (nozzle)
3	Quieter by design - air knives	Air dryer	General	Design (nozzle)/new blower
4	Flexible acoustic screening material	Axial fans	Mining/quarrying	Acoustic screen
5	Reducing impact noise from aluminium casks	Barrel impacts	Brewing	Isolation
6	Removing impactive strike	Barrel marking machine	Brewing	Design (impact reduction)
7	Reducing noise from a sugar beet cleaner loader	Beet cleaner loader	Agricultural	Design (isolation/new engine)
8	Reducing bottling line noise	Bottling line	Food	Enclosure
9	Improvements to a bumping machine	Bumping machine	Manufacturing	Maintenance/design
10	Enclosing a can filler	Can filling machine	Food	Enclosure
11	Acoustic refuges	Cardboard box production	Packaging	Refuge
12	Active noise control - pneumatic transport system	Centrifugal fans	Food	Active noise control
13	Reducing noise from a glass tempering line	Centrifugal fans	Glass	Enclosure (masonry)
14	Enclosing cold heading machines	Cold heading machine	Shoe	Enclosure
15	Removing jet noise	Compressed air hose	Pharmaceutical	Silencer (jet)
16	Reducing noise from a plastic mould cleaning gun	Compressed air tool	Engineering	Design (nozzle)
17	Reducing noise from screw compressors	Compressors	Water	Design (silencer)
18	Treating low-frequency compressor noise	Compressors	Manufacturing	Silencer
19	Enclosing ammonia compressors in seafood freezing	Compressors	Food	Enclosure
20	Acoustic lagging for a pneumatic conveying system	Conveying system pipes	Food	Lagging
21	Modifying an edge trimming machine	Edge trimming machine	Shoe	Absorbent lining
22	Active control of low-frequency, pure-tone noise	Engine	Automotive	Active noise control
23	Low-cost noise control of an engine test cell	Engine exhaust	Automotive	Resonator/silencer
24	Reducing noise in a dump truck	Engine/power pack	Construction	Absorptive/damping/lining
25	Reducing noise in a drag-line cab	Engine/power pack	Mining/quarrying	Absorptive/damping/lining
26	Reducing noise in crew cab road vehicles	Engine/power pack	Utility	Absorptive/damping/lining
27	Isolating a hydraulic guillotine	Hydraulic guillotine	Automotive	Isolation
28	Pneumatic impact press noise reduction	Impact press	Manufacturing	Isolation/silencer
29	Using a pneumatic squeeze press	Impact press	Engineering	Design (damping)/silencer
30	Machining alternator castings	Lathe	Automotive	Damping

Title	Noise source	Industry	Control method
31 Reducing noise when loading dump trucks	Material handling	Mining/quarrying	Damping
32 Material change - block making machine	Material handling	Brick	Design (new material)
33 Reducing impact noise from cocoa beans	Micronising machine	Food	Damping
34 Very high performance drywalling	Miscellaneous	Leisure	Partition wall
35 Enclosing a nail sorting machine	Nail sorting machine	Engineering	Enclosure
36 Reducing combustion noise by flow control	Oven	Food	Design
37 Quieter by design - paring machine	Paring machine	Shoe	Design (misc)
38 Reducing foundry pipe spinning noise	Pipe spinning	Steel/foundry	Isolation
39 Quieter by design - strand pelletisers	Plastic pelletiser	Plastics	Design (misc)
40 Acoustic enclosure of small plant	Pneumatic fan	Petrochemicals	Enclosure
41 Reducing noise from pneumatic screwdrivers	Pneumatic hand-tools	General	Design (isolation/duct exhaust)
42 Reducing air exhaust noise in vertical grinders	Pneumatic hand-tools	General	Design
43 Dynamic absorption of power press vibration	Power press	Automotive	Absorption
44 Shielding a printing press	Printing press	Printing	Acoustic screen
45 Reducing pump noise by good maintenance	Pump	Food	Maintenance
46 Use of absorption in noise control programme	Punch press	Pharmaceutical	Absorption (hung)
47 Flexible PVC enclosure of automatic punch presses	Punch press	Engineering	Enclosure (strip curtain)
48 High-speed press noise - modifying safety guarding	Punch press	Pharmaceutical	Enclosure (isolation/misc)
49 Anti-vibration treatment of high-speed presses	Punch press	Electrical	Isolation/damping
50 Contolling noise in a press shop	Reciprocating press	Engineering	Absorption (walls)/acoustic screen
51 Active/absorptive silencer for a rotary blower	Rotary blower	General	Hybrid active silencer
52 Reducing stone cutting noise	Saw blades	Stone cutting	Design (damping)
53 Reducing noise from extrusion line cut-off saws	Saw blades	Engineering	Enclosure
54 Fitting silencers to steam trap discharges	Steam discharge	Petrochemicals	Diffuser silencer
55 Silencing pressure vessel discharges	Steam discharge	Automotive	Silencer
56 Modifying acoustically-treated control booths	Steel rod production line	Steel/foundry	Refuge
57 Removing woodworking machine noise by adjustment	Thicknessing machine	Woodworking	Design
58 Reducing noise in a high-speed transfer press	Transfer press	Manufacturing	Design (impact reduction)
59 Rubber damping landing chutes	Unloading mechanism	Automotive	Damping
60 Damping compounds on heavy plates	Warehouse transport	General	Damping

1
Reducing noise in gravel chutes

The problem

A major noise source in quarrying and mineral workings is from materials dropping onto steel chutes or into hoppers. One sand and gravel company was faced with both high noise levels and a high plant wear rate on chutes at the end of a screening plant.

The solution

The chute was lined with a 25 mm thick, abrasion-resistant rubber. The material was attached to the steel surfaces of the chute using bolts and large diameter washers.

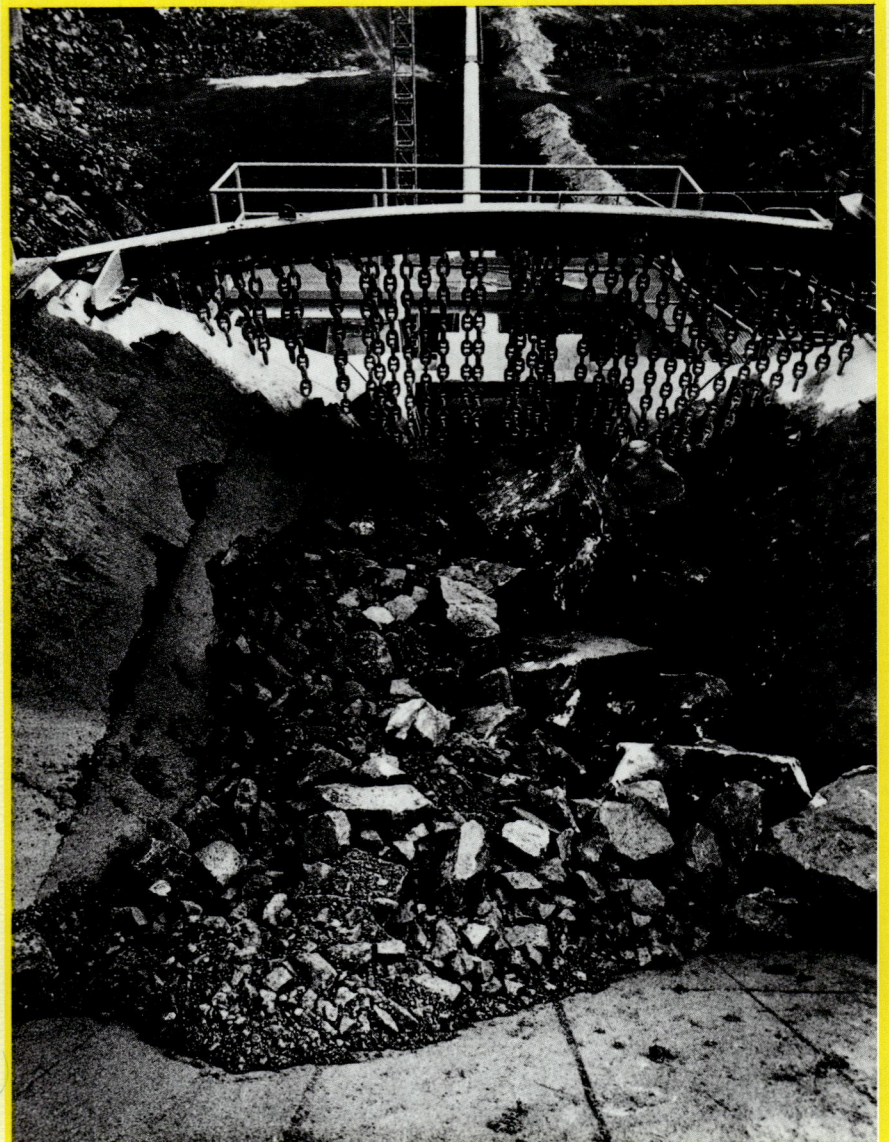

Gravel chute

The cost

About £2300.

The result

A noise reduction of 10 dB(A). Wear rates of the chute components were also reduced.

Information supplied by
Trellex Limited

The problem

Compressed air jets, widely used in industry for drying, can emit excessive levels of noise from the high jet discharge. One manufacturer measured noise levels of 96 dB(A) at operator positions coming from a bank of plain-ended, 6 mm diameter copper pipes being used to dry small components emerging from a washer on a conveyor line.

The solution

Noise reduction was achieved by replacing the plain jets with induced-flow type air knives. This followed the aerodynamic principle, known as the 'Coanda Effect', by which an air stream discharged at high velocity immediately next to a surface attaches itself to that surface and follows its contour.

In this case the contour was a rounded corner of a metal block. The effect of discharging the jet along one face was to produce a high velocity air flow along the other face at right angles to it. The air then left the second face to become a high-velocity free-stream flow directed at the washer conveyor line.

The primary flow was drawn along with a secondary flow from the still air through which it passed, so reducing the noise-generated turbulence. Primary air consumption was also reduced, resulting from the more efficient movement of secondary air.

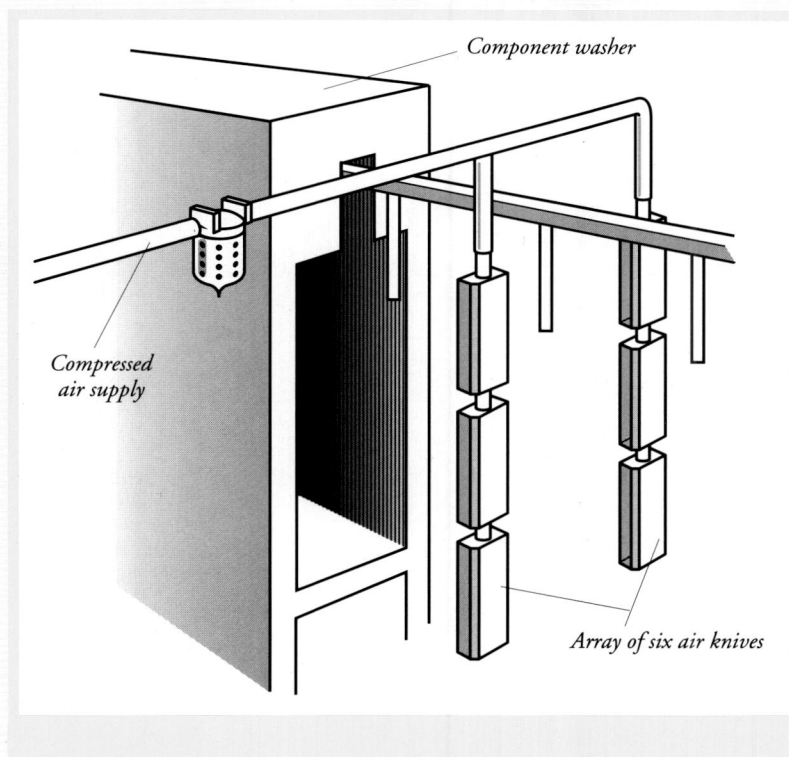

Component washer

Compressed air supply

Array of six air knives

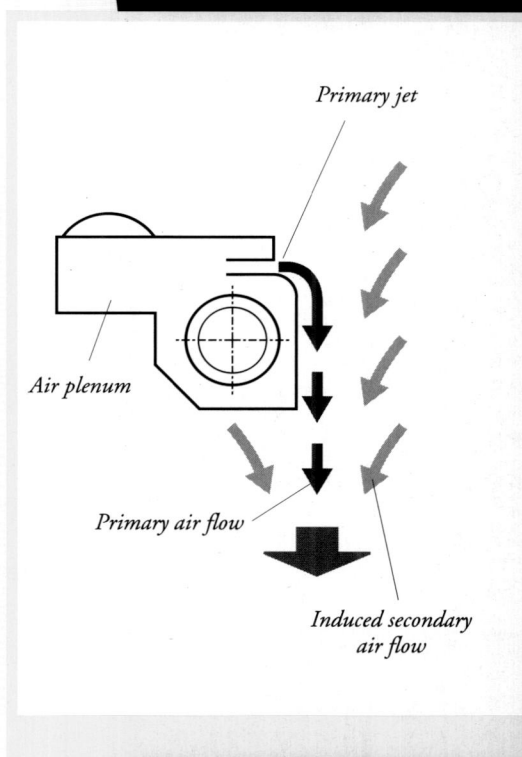

Primary jet

Air plenum

Primary air flow

Induced secondary air flow

The cost

About £900, plus installation.

The result

A noise reduction of 9 dB(A). The company has since reported a four week cost recovery on the installation and overall annual savings of some £6000 on the cost of compressed air production.

Report supplied by
Meech Exair

3
Quieter by design - air knives

The problem

Many manufacturing processes use air 'curtains' for drying, film control and product wiping. The air knife, for example, consists of a tubular plenum chamber stretching across the width of the product line, with a slot along its length. Air is supplied to the plenum at high pressure and leaves the slot as a high velocity 'blade' of air directed to strike the product line passing below.

Noise levels of between 90 and 95 dB(A) can be generated by a conventional multi-jet system supplied with air from belt-driven, single-stage blowers.

The solution

The use of a long continuous single jet rather than multiple individual jets can decrease velocity, thus reducing noise output from the jet itself. Jet noise can be further reduced if the air supply plenum is designed to have an aerodynamically smooth approach to the blade nozzle entry.

One manufacturer of this type of dryer used two direct-drive, multi-stage, centrifugal blowers for the air source to produce a flow of some 112 litres per second and a static pressure of up to 16 kilopascals (KPa).

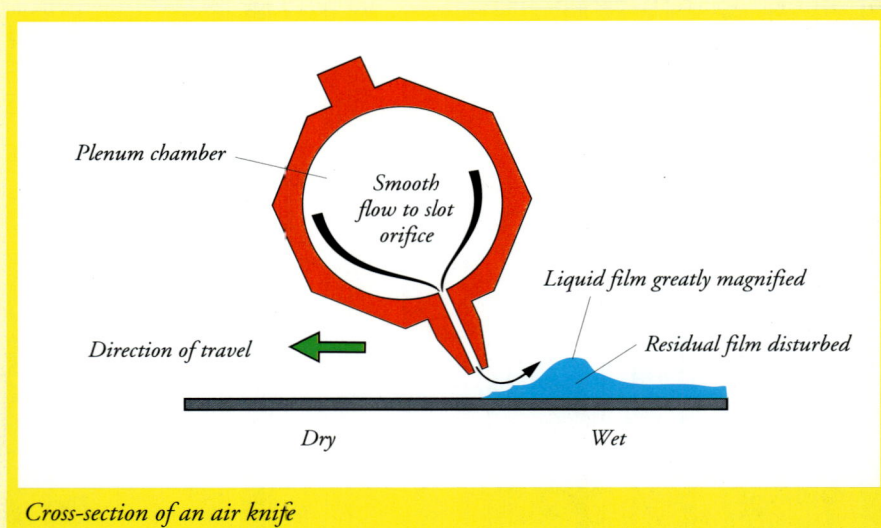

Plenum chamber

Smooth flow to slot orifice

Liquid film greatly magnified

Residual film disturbed

Direction of travel

Dry   *Wet*

Cross-section of an air knife

The cost

About £2000.

The result

A noise reduction of about 15 dB(A) at 3 m.

Equipment supplied by
Air Control Installations (Chard) Limited

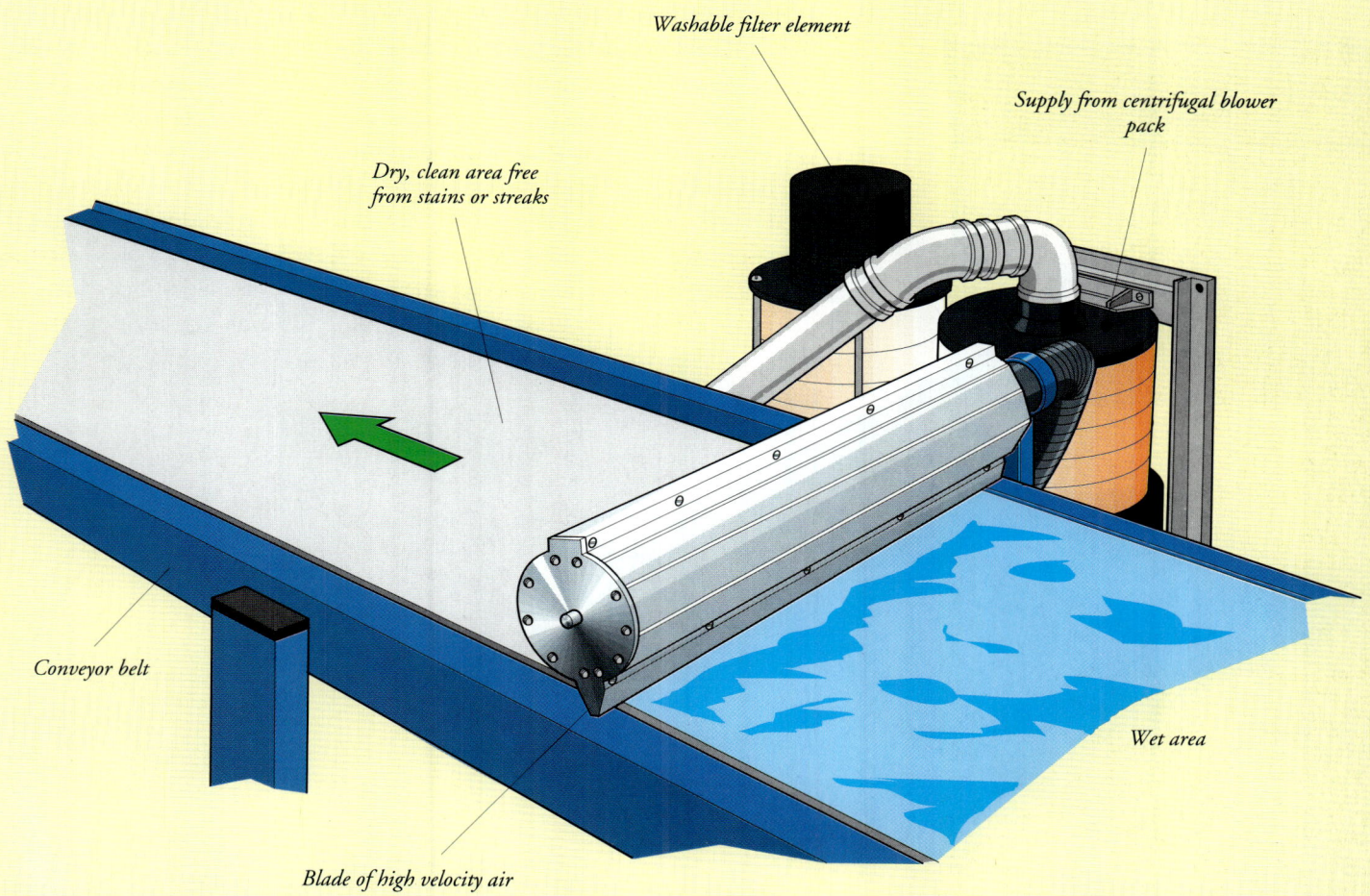

Washable filter element

Supply from centrifugal blower pack

Dry, clean area free from stains or streaks

Conveyor belt

Blade of high velocity air

Wet area

4
Flexible acoustic screening material

The problem

In mining and similar industries, high noise levels can be emitted from equipment such as pumps, local ventilation fans or hand-operated tools. Noise reduction can be complicated as these types of equipment have to be quickly relocated as work moves. In addition, conventional rigid acoustic enclosures may not fit into the confined work space.

The solution

In such cases, the only practicable alternative is to surround the source (and the operator of hand-held tools) with an acoustic screen. The material used for the screening has to be:

- heavy enough to provide a reasonable transmission loss;
- flexible enough to mould itself to the contours of any confined space;
- portable;
- capable of meeting underground safety requirements such as anti-static and fire retardancy;
- reasonably robust.

One material developed for this purpose consists of a 5 kg/m^2 thin lead sheet, bonded to a 25 mm thick glass fibre mat and encased in a heavy-duty, quilted plastic cover. The material can be made up as individual curtains with metal reinforced eyelets along the top edge and industrial grade 'Velcro' strips along the side edges. The acoustic screen can then be made up by hanging the curtains on the wire hooks on a pre-strung cable or free-standing frame, and sealing them together with the edge strips.

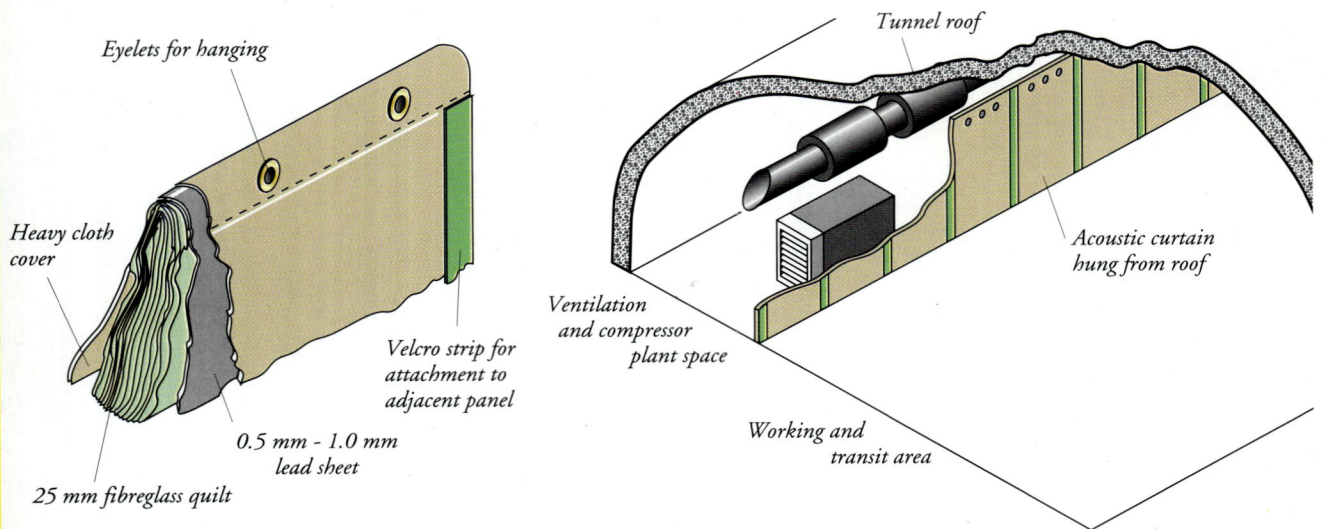

Eyelets for hanging

Heavy cloth cover

Velcro strip for attachment to adjacent panel

0.5 mm - 1.0 mm lead sheet

25 mm fibreglass quilt

Tunnel roof

Acoustic curtain hung from roof

Ventilation and compressor plant space

Working and transit area

The cost

£45 per square metre for the material.

The result

A noise reduction of 10 dB(A).

Information supplied by
The Noise Control Centre

The problem

Noise levels, typically of 93 dB(A), were being generated at a brewery by dropping aluminium casks onto the concrete surface of its distribution yard.

The obvious solution was to allow the casks to drop onto a resilient material, but this posed a number of practical problems. Individual movable mats were unsuitable, as 'drops' could take place anywhere in the yard and the mats would be difficult to move around. An entire covering of the yard with this resilient material, while possible, was discounted because the material would not be strong enough to withstand the punishing regime of repeated heavy blows, often with the sharp edges of the cask.

The chosen covering would have to resist being moved by road vehicles and site fork and pallet trucks, and be weather resistant. Fixing the material directly to the concrete with conventional adhesives was therefore impractical.

The solution

The solution was found by using a different type of resilient material and fixing it in place to cover the whole distribution yard. The material used was an abrasion-resistant rubber, pre-bonded to steel sheets that could be attached to the concrete using conventional masonry fixings.

Distribution yard

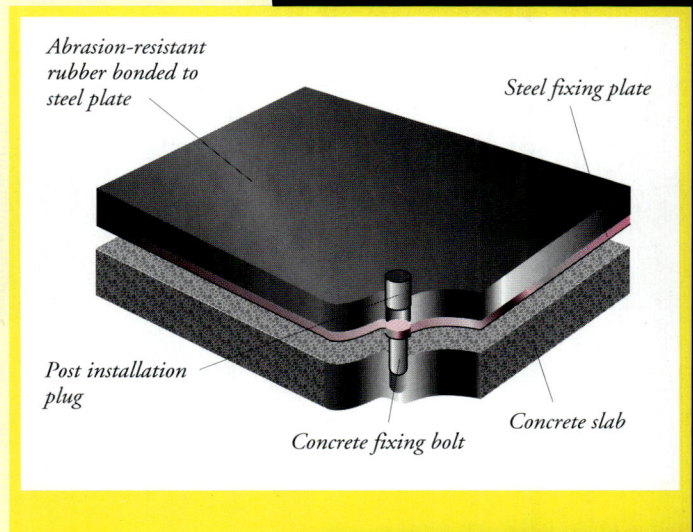

The cost

About £50 000.

The result

Peak noise levels were reduced by 10 dB(A). An additional benefit was improved maintenance and repair costs for the concrete, now protected from mechanical and chemical damage.

The material has been installed successfully in other industries, eg in premises handling gas cylinders and steel drums and in health clubs where weight training takes place.

Information supplied by
Trellex Limited

Photograph courtesy of
Bass Mitchells and Butlers

6
Removing impactive strike

The problem

Identification marking during the manufacture and refurbishment of beer barrels can result in noise levels of 105 dB(A). During marking, barrels are imprinted with a series of numbers and letters naming the owners and indicating the type and serial number of each batch.

Modern methods now use machines for large batches. Early designs used pneumatic impact cylinders to make an impression on the empty barrels. Acoustic screens limited noise radiation to neighbouring workers but did not reduce the exposure of the operators.

The solution

The new machinery combines pneumatics and hydraulics to control the impactive strike. The pneumatics accelerate the die towards the barrels but stop it short of the actual point of impact. The hydraulic compression finishes the impression stroke by applying a force of 15 tonnes over the last 3 mm of the die movement.

The cost

About £3000 per machine.

The result

A noise reduction of about 20 dB(A). An additional benefit has been an improvement in the quality of definition in the lettering and numbering.

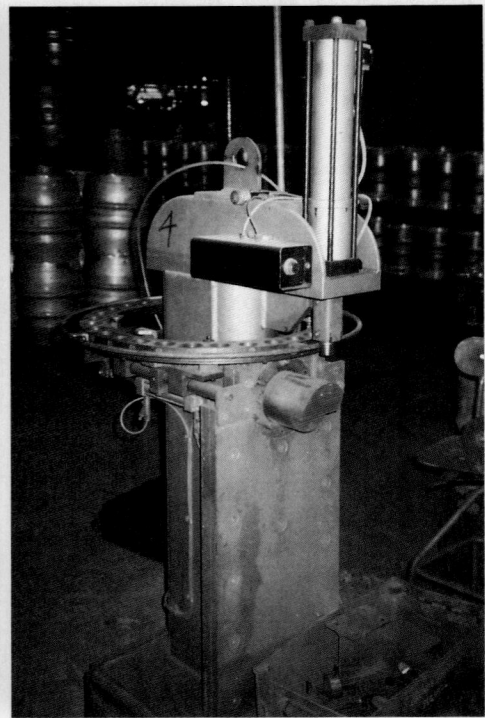

Barrel marking using pneumatics and hydraulics

Photographs courtesy of
Alumasc-Grundy Limited

The problem

In the agricultural sector, cleaner loaders raise the vegetable crop to a suitable height for loading on to a trailer, removing excess earth as the product is lifted.

In one company, sugar beet was carried on a conveyor web of metal bars, about 10 mm in diameter with 50 mm spacing between the bars. The ends of the bars were bent round to form a catch which linked to the next bar, resulting in a loosely linked steel chain. As the web rotated, the slackening and tightening of the different parts of the chain emitted noise levels of 103 dB(A) at the operating position.

Noise from an air-cooled 8 horsepower (hp) engine was also a problem, measuring 95 dB(A) while idling.

Old web arrangement

New web arrangement

The solution

The new design for the cleaner loader used the same rods, but with the ends embedded in a continuous rubber loop. This eliminated the impact of steel on steel in the web and at the driving gear.

Engine noise was reduced by replacing the old air-cooled machines with ones run by water-cooled 50 hp engines.

The cost

No extra cost for the new web arrangement.
Negligible difference between costs of the engines.

The result

A noise reduction of 9 dB(A) after the installation of the rubber web and a 7 dB(A) reduction in engine idling noise. Further reductions could be achieved by upgrading

Sugar beet cleaner loader

the engine silencer, or by using a power take-off drive from a tractor at low engine revs some distance from the operating position.

Equipment designed and manufactured by
Jones Engineering

8
Reducing bottling line noise

The problem

Bottling lines are a common source of high noise levels, posing particular problems for noise reduction. Noise is generated as the glass bottles strike against one other along the conveyors. Assuming that plastic bottles cannot be substituted for the glass, the only means of reducing the noise at source is to space the bottles to prevent collisions. Although possible along straight runs, most bottling lines require the marshalling of bottles at entrances to, or exits from, particular parts of the line such as washers, inspection tables or craters.

The only effective way to control the noise is to contain it within the line by acoustic enclosure or screening, which must be hygienic and allow immediate access to any section of the line.

The main bottling hall of a major food manufacturer was a highly reverberant area approximately 20 m long x 6 m wide x 5 m high, containing a total conveyor length of some 30 m, with noise levels above 85 dB(A).

Enclosure over bottling line

The solution

To reduce noise levels, an enclosure was erected over the conveyor. The roof was a polypropylene plastic sheet with an absorptive lining, faced with a thin plastic membrane to enable easy cleaning. The vertical side panels were in a transparent acrylic sheet, 13 mm thick, hinged at their top edges to allow operators rapid access. The side panels extended down to a few centimetres below the conveyor itself, but the area below was left open to enable excess conveyor lubricant to drain away.

The cost

£500 per metre of conveyor (1988).

The result

A noise reduction of about 5 dB(A). It was estimated that this could be increased to 10 dB(A) by extending the side panels to the floor with acoustic strip curtaining, thus containing downward radiated noise from the conveyor bed.

The problem

A manufacturer of artificial limbs was concerned with noise levels produced by a small bumping machine used for hand-forming sheet metal components of complex shapes. Unformed sheets were held over a steel ball and then pressed into a curve by an elastic pad on the end of an arm. The arm was repeatedly and rapidly raised and lowered by a mechanical drive, resulting in noise levels in the range 91 to 93 dB(A).

The noise generating mechanisms were identified as the rigidly-mounted drive assembly, badly worn bearings, excessive clearances and the drive-belt guard.

The solution

The following modifications were made to the machine:
- replacing the sheet metal drive-belt guard with wire mesh;
- using bobbin mounts to isolate the drive motor bed plate from the main casting;
- cutting holes in the bed plate to reduce its area and hence noise radiation;
- pre-loading the drive mechanism to reduce rattle;
- replacing plain bronze bearings with accurately loaded taper roller bearings;
- reforming the steel anvil ball socket for a tighter fit.

The cost

About £800.

The result

A noise reduction of between 16 and 21 dB(A), depending on whether the machine was idling or in full operation. An additional benefit was to extend the life of the bearings from six months to three years.

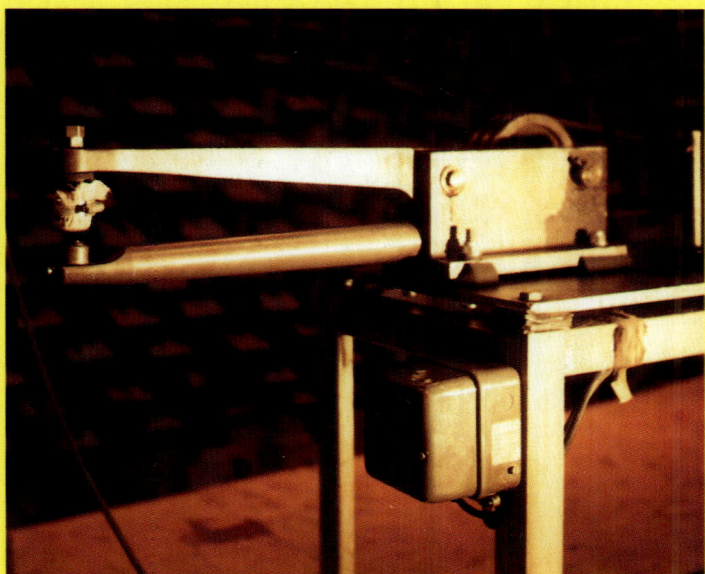

Bumping machine

Report supplied by
ISVR Consultancy Services

10
Enclosing a can filler

The problem

A soft drinks can filling line, operated by a large food manufacturer, contained three closely-spaced machines, a can filler, an aeration unit and a lid seamer for closing the filled can. The machines were all situated in a reverberant room where noise levels of 99 dB(A) were measured.

The solution

Noise reduction was achieved by fitting the seamer with an acoustic enclosure, specifically designed to meet hygiene and access criteria.

The basic panel work of the enclosure consisted of polypropylene and 13 mm thick ME grade perspex sheet (ie suitable for enclosures and guards), chosen for their durability under repeated chemical cleaning. Where visual inspection was unnecessary, the panels were lined with 25 mm, acoustically absorptive foam, wrapped in an impervious film to stop oil and moisture getting in. All doors were provided with safety interlocks to prevent access while the seamer was in operation.

It was necessary to leave infeed and discharge areas of the enclosure open, for example to clean the seamer and to clear jams. This inevitably reduced overall enclosure performance. However, the directional high-frequency character of the noise could be blocked from the main source areas.

The cost

About £5000 (1988).

The result

A noise reduction from the seamer of 17 dB(A).

Filled can infeed conveyor

Can lid infeed

Information supplied by
SmithKline Beecham Consumer Healthcare

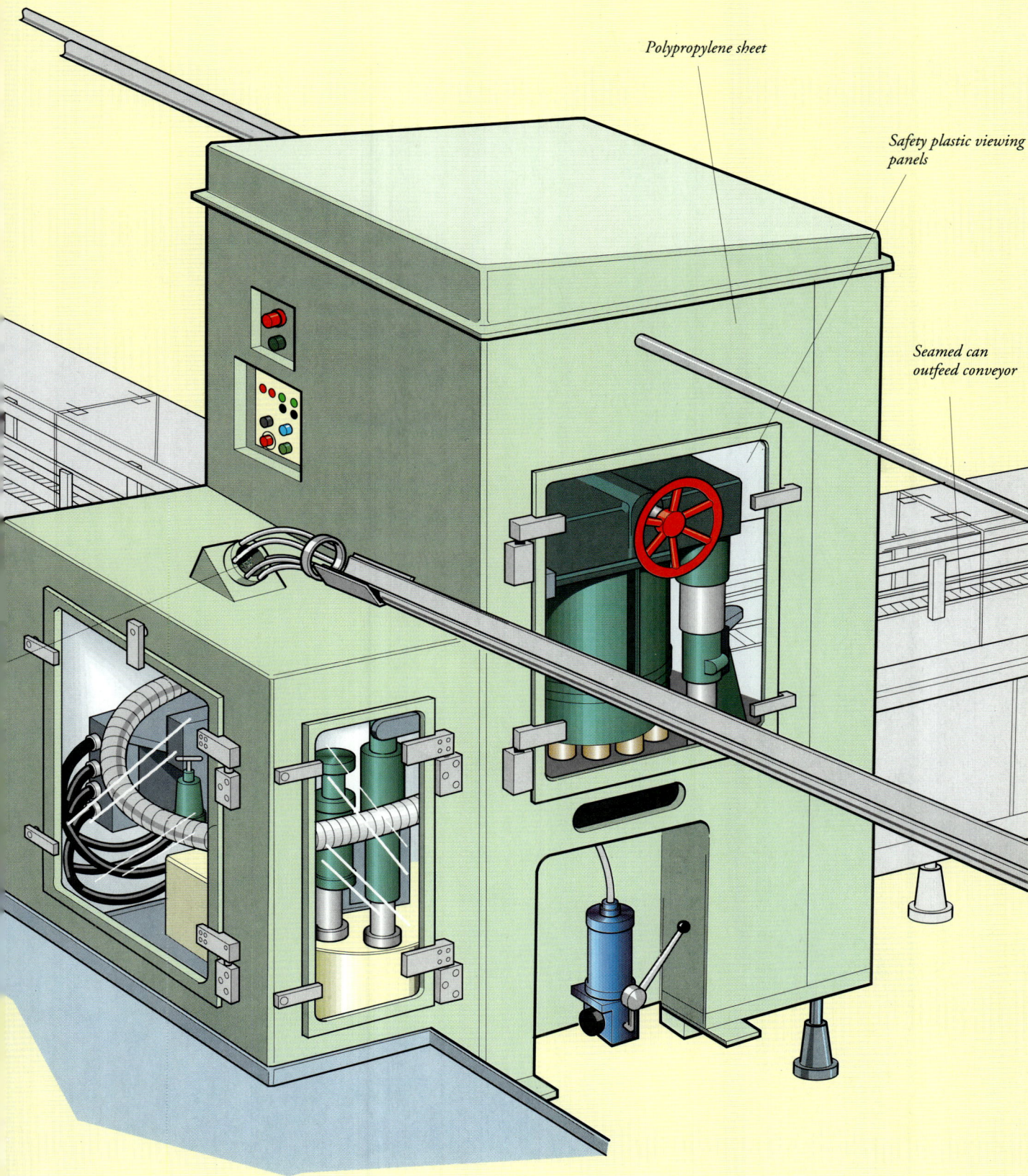

Polypropylene sheet

Safety plastic viewing panels

Seamed can outfeed conveyor

Can filling enclosure

19

11
Acoustic refuges

The problem

Where it is impossible to separate workers from noisy equipment, but their hearing needs to be protected, an alternative approach is to use noise refuges. For example, a firm that manufactured cardboard containers installed a noise refuge to provide an effective working area close to the main production lines where noise levels approaching 97 dB(A) had been measured.

The solution

The firm had a fully air-conditioned enclosure installed, consisting of double-skin acoustic panels that incorporated acoustic glazing to give good all round visibility into the production areas. Internally, the refuge was finished to office standards - tiled floors and acoustically absorbent ceiling and walls.

Acoustic refuge in production area

The cost

No costs available.

The result

The example shown produced a difference between outside and inside noise levels of some 30 dB(A).

Refuge manufactured and supplied by
Noise Reduction Limited

The problem

Centrifugal fans are often used to convey material around production lines. In one drinks canning plant, a six-bladed radial fan in the filler room, delivering 3 m³/s of air, supplied air to move cans along lines. The fan produced a tone at about 300 Hz with an overall noise level of over 90 dB(A).

The solution

It was decided to adopt an active noise control solution. This consisted of a digital signal processing controller producing a digital model of the noise. A signal microphone fed information to the controller which then produced an inverse sound wave from the loudspeaker. This 180° out-of-phase wave combined with the noise, cancelling it by destructive interference. A second microphone analysed the operation of the system and adapted it continuously to take account of system changes.

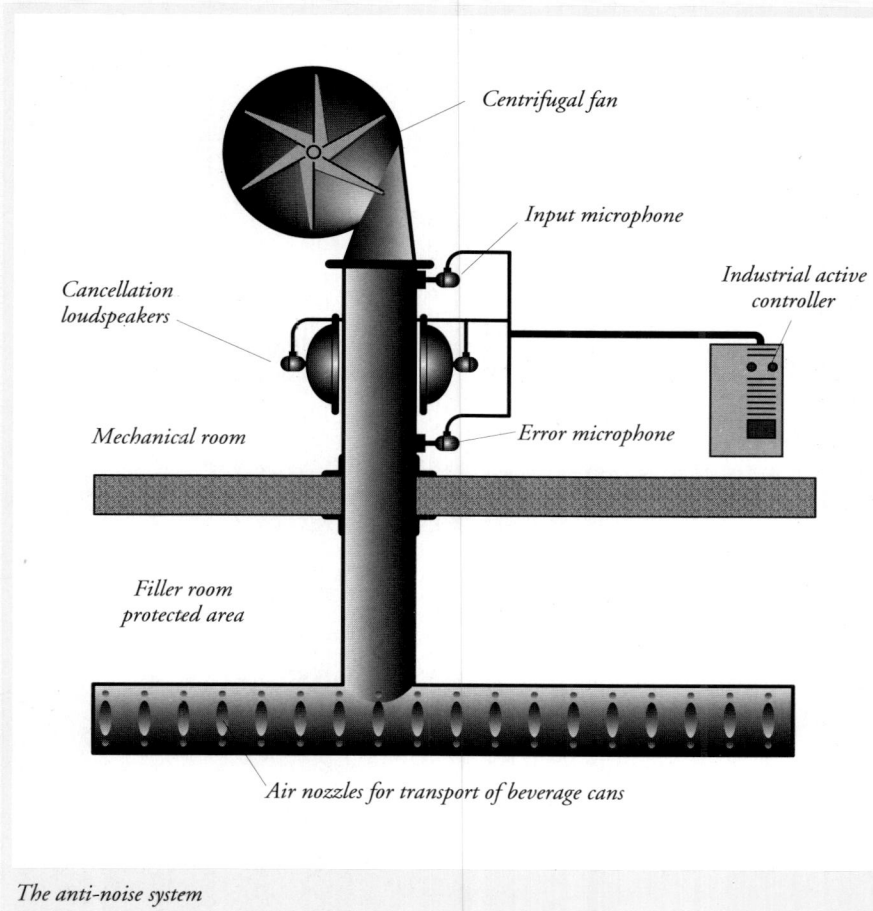

Centrifugal fan

Input microphone

Industrial active controller

Cancellation loudspeakers

Mechanical room

Error microphone

Filler room protected area

Air nozzles for transport of beverage cans

The anti-noise system

The cost

£6500.

The result

A reduction in the tonal noise by 22 dB. There was no interference with the work process, such as might occur with an enclosure.

Equipment manufactured by
Digisonix Inc, USA

13
Reducing noise from a glass tempering line

The problem

Glass tempering is achieved by heating glass and then quickly cooling it at a controlled rate to align the molecules in one direction. Air is used to cool the glass: the thinner the glass, the greater the quantity of air required. In this case the company was producing 4 mm thick glass and the main cooler fan was delivering some 21 m³/s.

Traditionally, cooler fans are mounted outside the factory to reduce the impact of the fan noise on the workers. However, in this plant, the fans had to be mounted internally. With fan sound power levels in excess of 126 dB(A), the effective acoustic enclosure of the fans was essential.

The solution

To minimise costs, a masonry enclosure, rather than one with conventional metal acoustic panels, was constructed. The enclosure consisted of 225 mm thick, heavy-density blockwork (2400 kg/m³) walls and a 250 mm deep timber joist roof. The space between the joists was packed with 100 kg/m³ mineral wool. A 50 mm thickness of plaster board was fixed to the top and bottom, with staggered and taped joints. Access to the fan room was via two large double door sets (3 m wide x 3 m high).

The very high volume of intake air was brought in through vertically-mounted rectangular duct attenuators (1.2 m wide x 2 m high x 2.4 m long). The thickness of the silencers within the attenuators was tuned specifically to give enhanced low-frequency attenuation without adding excessive pressure drop to the fans.

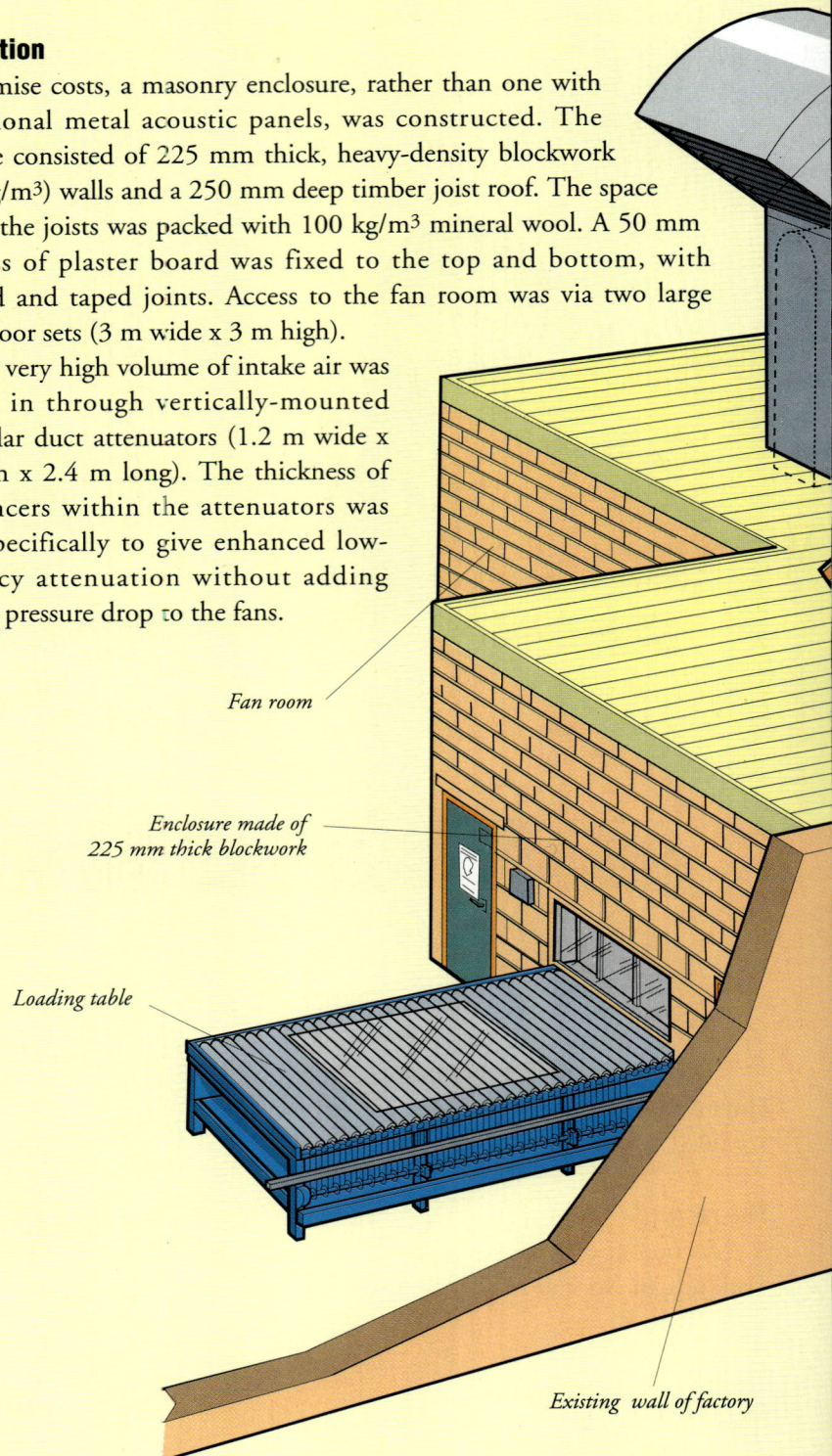

Fan room

Enclosure made of 225 mm thick blockwork

Loading table

Existing wall of factory

Information supplied by
Preedy Glass Limited

Tempering line supplied by
EFCO Limited

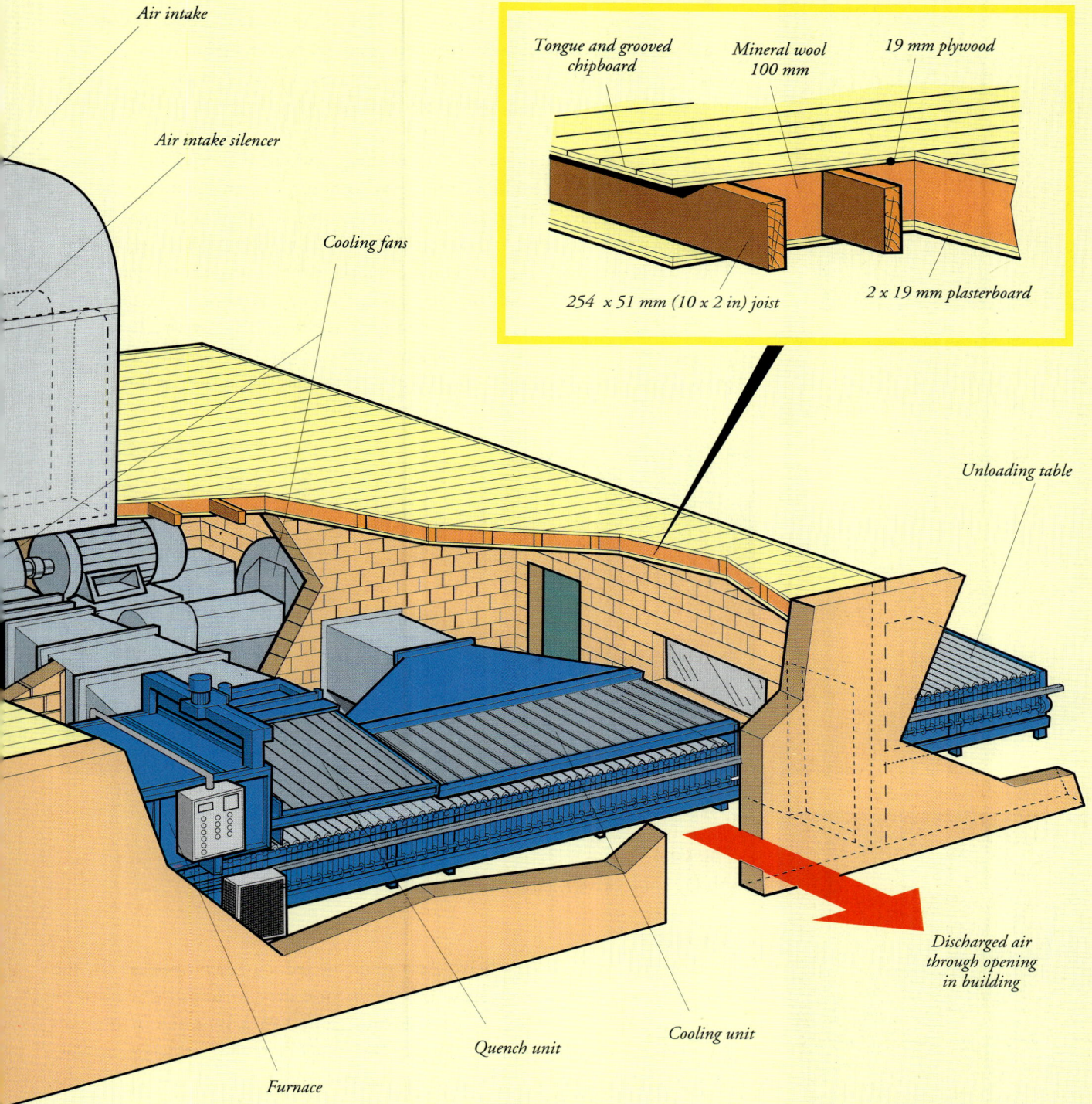

Air intake

Air intake silencer

Cooling fans

Tongue and grooved chipboard

Mineral wool 100 mm

19 mm plywood

254 x 51 mm (10 x 2 in) joist

2 x 19 mm plasterboard

Unloading table

Discharged air through opening in building

Furnace

Quench unit

Cooling unit

The cost

About £15 000. The cost of the project was thought to offer a 50% saving over an all-steel acoustic enclosure.

The result

A noise reduction of about 45 dB(A).

14
Enclosing cold heading machines

The problem

The factory floor of one shoe manufacturer contained a considerable number of similar cold heading machines, with noise levels measured at around 104 dB(A).

The solution

The company approached an acoustic consultant to help design a custom-made enclosure. This included features required by the operators, such as an easy to raise hood and profiled corners for easy access. The design was first built in plywood to check on its acceptability. Once approved, tenders were sought for a steel version.

After receiving quotations, the company realised that it could achieve a significant saving by building the units in-house, in conjunction with the acoustic consultant. The enclosure was built using standard acoustic panels of 50 mm thick mineral wool between skins of plain and perforated steel sheet. The panels also included a Melinex wrapping to protect the mineral wool from oil and an anti-drumming sheet to limit panel vibration.

The enclosures had three openings - the wire feed at the front that passed through a tube surrounded by a brush strip; the product discharge chute; and a drive belt fed through a PVC plate with two slots cut into it.

Enclosed cold heading machines

The cost

About £400 per enclosure.

The result

A noise reduction of 19 dB(A). As a result of reducing the noise levels, the company was able to reorganise the production area to improve work flow and response times. Machines were positioned in a manufacturing cell arrangement together with equipment for subsequent processing operations, inspection and packing. The benefits of the reorganisation meant that the costs of the noise control programme were recouped after only three years.

Photographs courtesy of
British United Shoe Machinery Limited

Acoustic consultants were
R S Allsopp and Associates

The problem

During its production process, a health care company immersed its pharmaceutical containers in a rinsing bath. Afterwards, to dry off excess moisture, employees placed the containers on a draining bench and blew them dry using a hand-held compressed air hose.

The solution

The plain air hose nozzle was replaced by an induced flow type, the principle being to link a secondary air flow with the primary jet flow and so reduce the strength of the noise-generating turbulence produced in the main jet air flow.

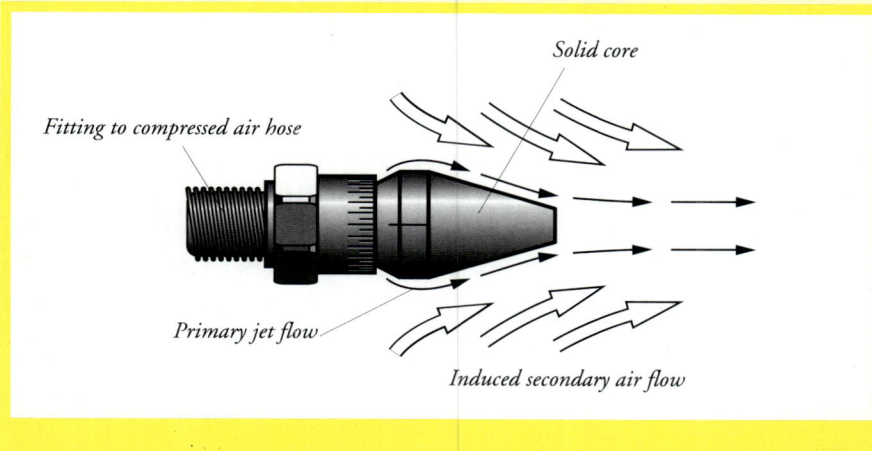

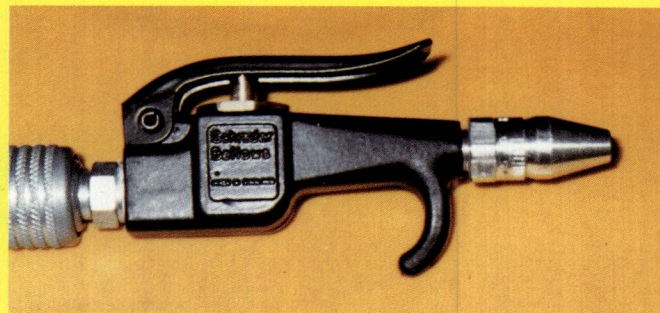

Solid core

Fitting to compressed air hose

Primary jet flow

Induced secondary air flow

Compressed air hoses

The cost

About £25 per unit.

The result

A noise reduction of 7 dB(A) during the drying cycle, while still maintaining an effective velocity for moisture removal.

Information supplied by
3M Health Care

Photographs courtesy of
Meech Exair

Equipment supplied by
Meech Exair

16
Reducing noise from a plastic mould cleaning gun

The problem

A company engaged in hydraulically moulding thermosetting plastic components used a compressed air gun to clear flash and powder deposits from the die moulds. Before it was redesigned, the gun consisted of a plain steel tube, with an internal diameter of approximately 1/8 in, screwed into the end of a compressed air hose and directed at the appropriate area. The interaction of the high-velocity jet and the surface of the moulds produced noise levels reaching 105 dB(A).

The solution

The plain nozzle was replaced by one generating an induced secondary air flow through an aspirated venturi. The primary flow was passed through a venturi nozzle inside a tube, drilled in its sides to allow outside, secondary air to pass through the wall and mix with the primary jet at the venturi discharge. The effect was two-fold:

- the mix of primary and secondary flows within the tube reduced the amount of turbulence and hence the sound generated at the primary nozzle;
- the larger diameter of the tube itself reduced air speed onto the mould dies, even allowing for the increased volume of the discharge air, and hence reduced noise generation at the die mould surfaces.

The cost

About £40 per nozzle.

The result

A noise reduction of up to 10 dB(A). Another benefit was increased mechanical strength of the tube which allowed the occasional flash or powder deposit on the die to be removed by tapping or scraping with the tube edge.

Information supplied by
Industrial Noise Services Limited

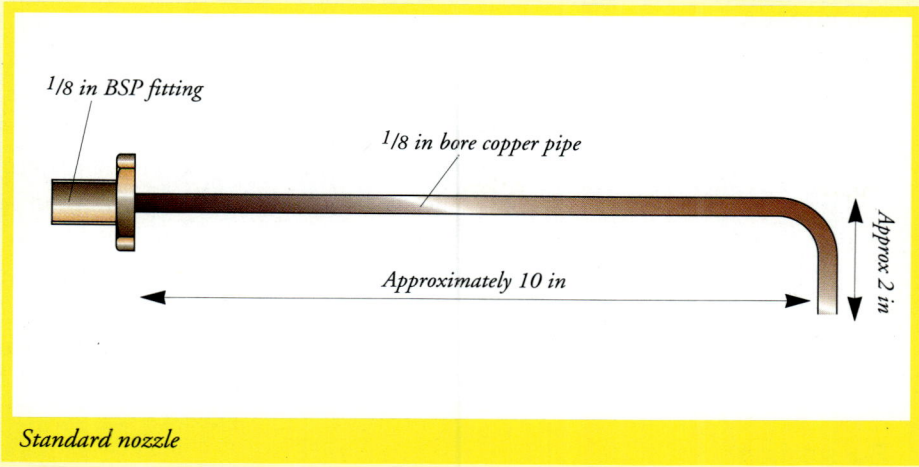

¹/8 in BSP fitting

¹/8 in bore copper pipe

Approx 2 in

Approximately 10 in

Standard nozzle

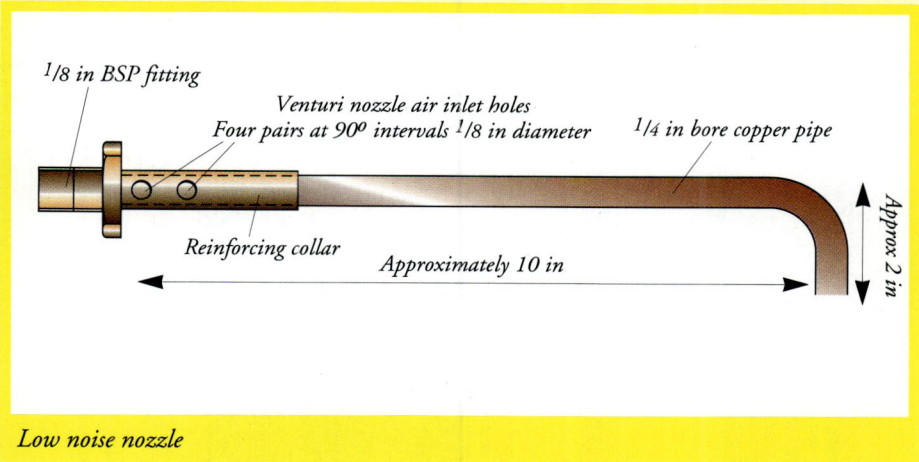

¹/8 in BSP fitting

Venturi nozzle air inlet holes
Four pairs at 90⁰ intervals ¹/8 in diameter

¹/4 in bore copper pipe

Reinforcing collar

Approximately 10 in

Approx 2 in

Low noise nozzle

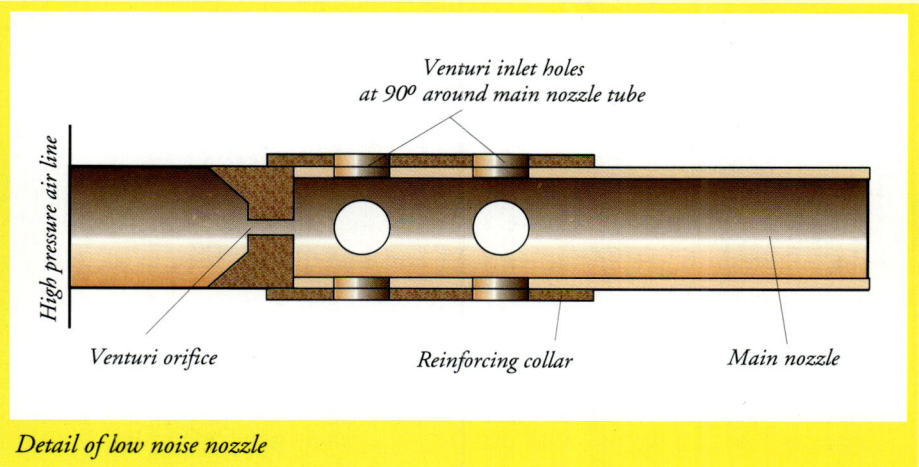

Venturi inlet holes
at 90⁰ around main nozzle tube

High pressure air line

Venturi orifice

Reinforcing collar

Main nozzle

Detail of low noise nozzle

The problem

Screw compressors providing an air supply for the aeration tanks in the blower house of a sewage treatment works were exposing maintenance engineers to noise levels of about 100 dB(A).

Although acoustic enclosures over each of the three blowers were effectively containing casing and drive noise, the air intake was being piped in from a filter outside the enclosure. Therefore noise generated inside the blower itself was passing into the inlet pipework and then escaping into the blower house.

The noise had a strong tonal content appearing with a peak at the lower end of the frequency range, in this case in the 63 Hz octave band.

The solution

As conventional absorptive silencers were inefficient at these frequencies and disproportionately large, the solution was to fit a reactive silencer into each air intake pipe. The strong noise components in the 63 and 125 Hz octave band, indicating strong harmonic content, meant that a twin-chamber design was required.

The chambers were arranged in series with the first designed to resonate, giving maximum attenuation at the fundamental frequency. The second was designed to reduce energy at the principal harmonic frequencies.

Screw compressors with reactive silencers

The cost

£900 per silencer.

The result

There was an overall reduction in the noise level of 20 dB(A); a 37 dB reduction in the 63 Hz frequency band; and a 30 dB reduction in the 125 Hz frequency band.

Sound pressure level at 1 m from the compressor intake

Frequency (Hz)	dB(A)	63	125	250	500	1000	2000	4000	8000
Before treatment	100	120	114	99	88	83	74	69	67
After treatment	80	83	84	84	74	71	70	68	62
Attenuation (dB)	20	37	30	15	14	12	4	1	5

Photograph courtesy of
Thames Water Utilities Limited

Equipment designed and supplied by
Ian Sharland Limited

The problem

Large reciprocating air compressors producing high noise levels at low frequencies can be found in a wide range of manufacturing plants. In one company, two compressors operating at 370 rpm (revs per minute) drew air from the roof where the intake filters and other items of equipment needed regular attention. The noise level near the air intakes was found to be 88 dB(A), determined almost entirely by energy in the 63 and 125 Hz frequency bands.

Absorptive silencing was discounted as inefficient at these frequencies. It would also have been impractical as the size of silencer necessary would have meant a need for additional roof support.

The solution

The solution was to install an intake silencer that could combine adequate low-frequency attenuation with low resistance to the air flow.

A combined reactive and absorptive silencer was selected for each compressor. The silencers consisted of an all-welded outer tube with a cross-sectional area at least 15 times greater than that of the system side pipework. This change in cross-sectional area resulted in a loss of sound energy due to wave reflection and cancellation. An acoustically absorptive lining within the enlarged silencer casing further reduced the noise level.

The cost

About £2500 each.

The result

The attenuation achieved in this example with a 200 mm diameter intake pipe and an 800 mm outside diameter silencer casing with an overall length of 3000 mm was sufficient to render the compressor noise effectively inaudible above other sources in the working area. The residual noise of adjacent equipment masked the full performance of the attenuator at the higher frequencies.

Air compressor with combined reactive and absorptive silencer

The sound pressure levels are shown in the table, along with the actual octave band sound reduction levels. Overall there was a 17 dB(A) reduction and an attenuation of 35 dB at 31.5 Hz.

Sound pressure level at 1 m from the compressor intake

Frequency (Hz)	dB(A)	31.5	63	125	250	500	1000	2000	4000
Before treatment	88	115	111	98	77	77	73	69	66
After treatment	71	80	80	76	66	68	65	58	56
Attenuation (dB)	17	35	31	22	11	9	8	11	10

19
Enclosing ammonia compressors in seafood freezing

The problem

Refrigerant compressors are a common source of noise in the food processing industry. Sometimes they have to be located close to the production facility served, occasionally in occupied areas.

A major seafood producer installing freezing lines had to site four 250 kW liquid ammonia compressors in a product transfer area regularly used by production personnel. Modifying the compressor to control the noise at source was not possible, leaving acoustic enclosure of the installation as the only practicable alternative.

The design of the enclosure had to provide rapid access to the compressors for maintenance. In addition, because of the nature and size of the machines, a continuous throughput of air was vital to remove the heat radiated from the compressors and their drive motors, and to prevent the build-up of any leaked ammonia gas inside the enclosure.

Cut-away of an acoustic room for refrigeration compressors

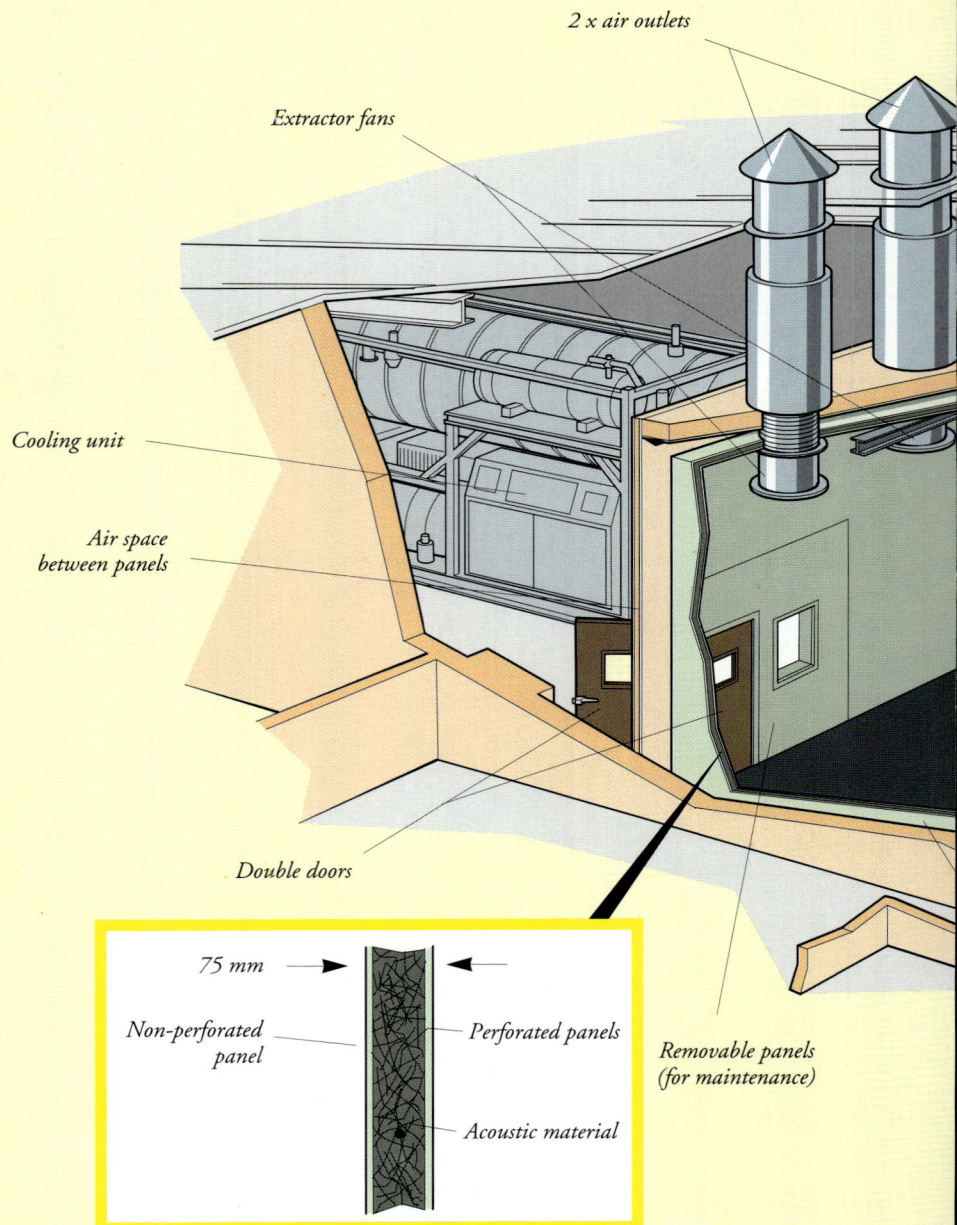

- 2 x air outlets
- Extractor fans
- Cooling unit
- Air space between panels
- Double doors
- 75 mm
- Non-perforated panel
- Perforated panels
- Acoustic material
- Removable panels (for maintenance)

Information supplied by
Lyons Seafoods Limited

Equipment manufactured and installed by
Ian Sharland Limited

The solution

The enclosure was constructed in acoustic panelling made up of a plain sheet steel outer skin and a perforated steel inner skin. Mineral wool slabs were inserted into the 75 mm wide space between the skins. The outer skins were formed in plastic coated steel to meet hygiene requirements that demanded a corrosion-free, easy-to-clean external surface. The panels were supported on a separate structural frame, free of the enclosed machinery to minimise any reduction of acoustic performance due to machine induced vibration. The overall size of the enclosure was 11 m long x 5.5 m wide x 4.5 m high.

The cooling air supply was provided by four fans each moving some 6 m³/s of air through the enclosure, two drawing in fresh air via an attenuated inlet duct, and two extracting air from the enclosure through a similarly treated discharge duct.

The cost

£45 000 installed.

The result

A noise reduction of at least 24 dB(A).

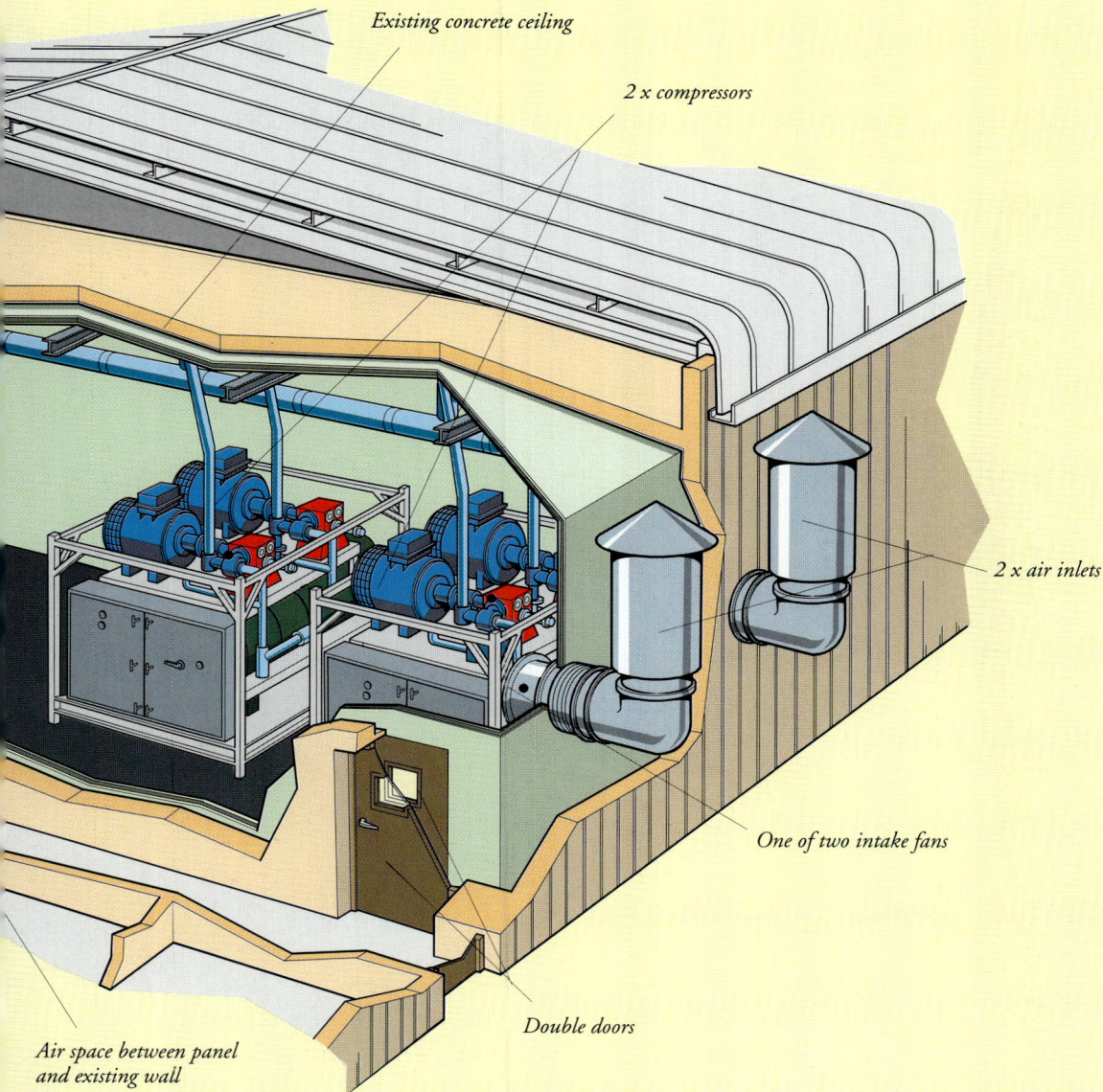

Existing concrete ceiling

2 x compressors

2 x air inlets

One of two intake fans

Double doors

Air space between panel and existing wall

20
Acoustic lagging for a pneumatic conveying system

The problem

Conveying solid particle products through rigid pipe systems is an inherently noisy process as the particles hit the undamped metal pipe walls.

Few available materials exist that can cushion the impacts by incorporating internal 'soft' linings. Many need to withstand the effects of long-term abrasion; others have to meet strict hygiene requirements. There are also engineering difficulties when fixing linings to often intricate runs of small diameter piping.

The solution

One flour mill reduced noise levels by applying acoustic lagging to the external surface of the pipe run. It was also necessary to ensure rapid access to a number of inspection panels built into the pipe runs and for the lagging to be removable for pipe cleaning.

Soft thermal-type lagging was discounted on mechanical grounds and because it was difficult to handle. Instead the lagging was pre-fabricated in a series of split semi-circular sections. Each section consisted of a half cylinder of plain sheet steel lined with 50 mm thick semi-rigid mineral wool slab, retained in the half cylinder by perforated sheet steel. The inner diameter of the section was equal to the outer diameter of the conveying pipe concerned.

Two matching sections were placed on opposite sides of the pipe and held in place by over-centre, quick-release toggle clamps. In some cases it was acoustically acceptable to leave one access door in the pipe unlagged. However, where a door was radiating significant noise, the internal perforated sheet and mineral wool infill were omitted from the lagging section and a corresponding hinged access door was provided in the external lagging shell.

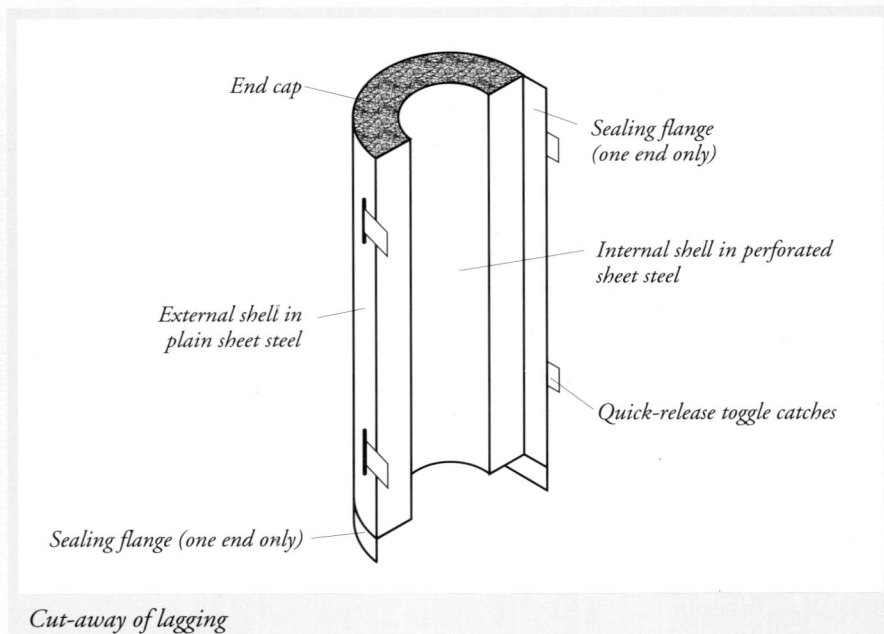

End cap

Sealing flange (one end only)

Internal shell in perforated sheet steel

External shell in plain sheet steel

Quick-release toggle catches

Sealing flange (one end only)

Cut-away of lagging

The cost

About £150 per metre lagged.

The result

A noise reduction of about 10 to 15 dB(A). This system of pipe lagging was found to be easily adapted to even complex pipe runs including bends and junctions.

Information supplied by
Ian Sharland Limited

The problem

In shoe manufacture, sole forming leaves excess material at the edges which must be removed by holding each shoe against a high-speed rotating blade.

Machines traditionally designed for this purpose consist of a cutting head located inside a five-sided metal box. Operators get at the blades through an open sixth side, with the off-cut flashings either falling to the floor or being removed by a dust extraction system at the rear. The open front of the machine gives little acoustic protection for the operators and noise levels can exceed 90 dB(A). The main source is thought to be the central spindle rotating at high speeds, generating a high-frequency tone.

The solution

One company fitted a simple system to each of their edge trimming machines by lining the inside of the machine casing with a 50 mm thick foam slab. This absorption restricted the build-up of reverberant sound energy inside the machine.

Each machine was also fitted with a 'half-height' clear polycarbonate plastic screen at the front, hinged along the top edge to enable it to be lifted clear, giving full access to the space inside. During operation, the screen could be lowered to give partial cover at the front, leaving a gap for the operator to reach in with the shoe. The treatment has proved to be a good compromise between screening direct sound radiation and maintaining efficient and safe operation.

The cost

About £50 per machine.

The result

A noise reduction of about 5 dB(A).

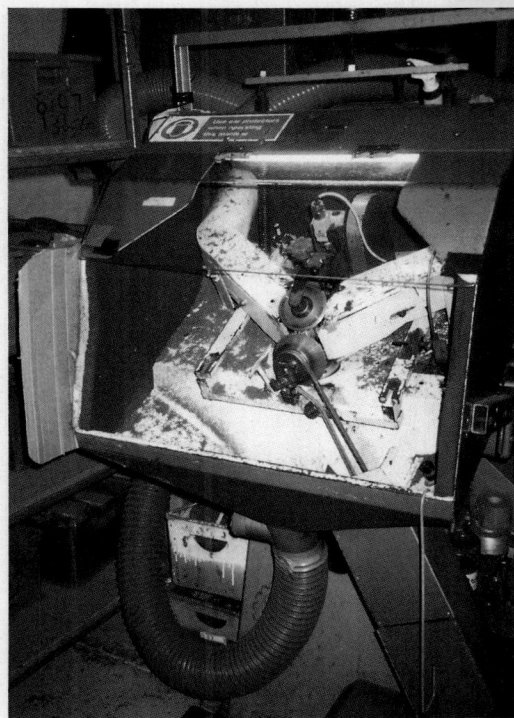

Edge trimming machine

Photograph courtesy of
Clarks International

22
Active control of low-frequency, pure-tone noise

The problem

Sound pressure can be reduced in some cases by active noise control, sometimes known as 'anti-noise'. This technique uses the principle of sound cancellation. The characteristics of a sound field in a given environment are measured and then reproduced by means of an artificial source such as an array of loudspeakers. The reproduced sound is generated in such a way that the new sound field produces a wave of, for example, positive pressure in exactly the same position and at exactly the same time as a negative pressure wave produced in the original sound field. The original sound wave is thus cancelled out or at least significantly reduced.

In practice, effective noise reduction by this technique is possible only for low-frequency pure tones occurring within well-defined enclosed spaces. Most attention has so far been devoted to the active control of noise in ducted systems carrying noise from such sources as fans and engine exhausts, resulting in some successful solutions.

The solution

More recently, attention has turned to the possibility of active noise control in larger spaces subjected to low-frequency, pure-tone noise. One example was the suppression of engine-induced 'boom' inside cars. Since this type of noise is primarily low frequency, conventional methods of acoustic insulation, besides adding both cost and weight to the vehicle, are not very effective. However, active noise control can provide a viable solution. By sampling the engine-generated sound field at a number of positions inside the vehicle, and by reproducing the field with the appropriate phase shifts using the vehicle's own loudspeaker system, noise reductions can be achieved at some positions in the vehicle.

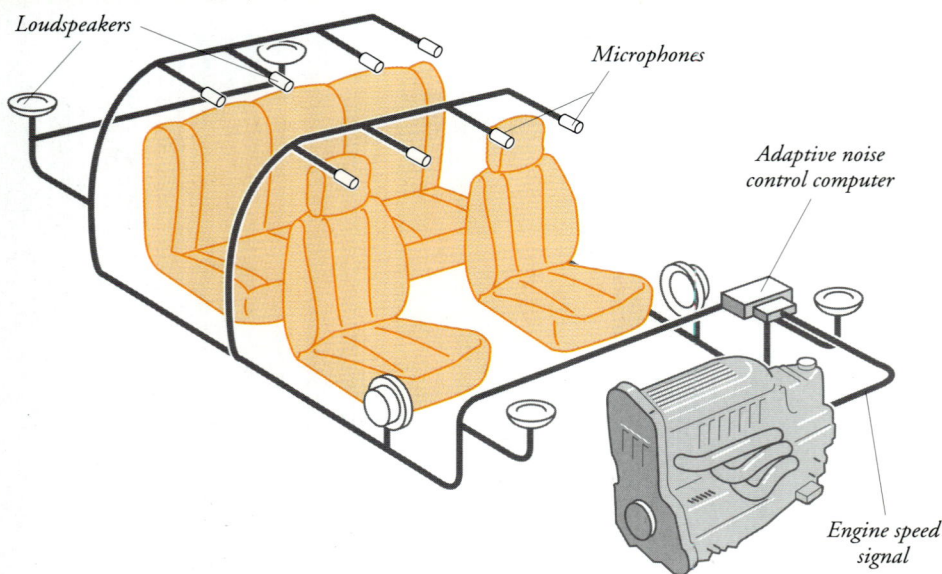

Schematic anti-noise layout inside a passenger compartment

Labels: Loudspeakers, Microphones, Adaptive noise control computer, Engine speed signal

The cost

£2000 for equipment, plus project-specific development and programming costs.

The result

Reductions of up to 10 dB in individual frequency bands and 5 dB(A) overall. There may be similar benefits for operators of commercial vehicles and industrial site and construction vehicles.

Low-frequency, pure-tone noise generated in small production areas by, for example, vibratory machinery or reciprocating air compressors may have characteristics appropriate for active noise reduction measures.

Report reproduced courtesy of
Dr P Nelson and **Dr S Elliott**,
ISVR, University of Southampton

The problem

In one company, an engine test cell had been generating high noise levels in maintenance areas near the exhaust discharge. The engine exhaust stack ran from ground level to the test cell roof where it ended in a pair of reactive and absorptive silencers.

Preliminary measurements revealed that the strongest tonal components appeared at 289 Hz with a smaller peak at 144 Hz. The major noise frequency was found to correspond with the engine firing frequency. For the straight six cylinder engine under test, at 5800 rpm, the fundamental firing frequency could be calculated at 290 Hz, with a subharmonic at 145 Hz.

The solution

As the tone fell between the optimum performance of the reactive and absorptive silencers, a custom-designed silencer was required. A tuned side-branch (Helmholtz) resonator, 'teed' off the main outlet, was chosen. The length of the side branch (29 cm) was one quarter of the wavelength at the critical frequency. The length of the silencer could be changed in situ to compensate either for manufacturing tolerances or for any slight shifts in the tonal frequency.

The silencer was made in-house from a section of 3 in (76 mm) mild steel steam pipe. The internal surface of the pipe was left plain as a particular tone was to be reduced - it was expected that an absorbent lining would have broadened the range of the silencer's performance, but lowered its attenuation at the specific frequency of interest.

The cost

£20 (1989).

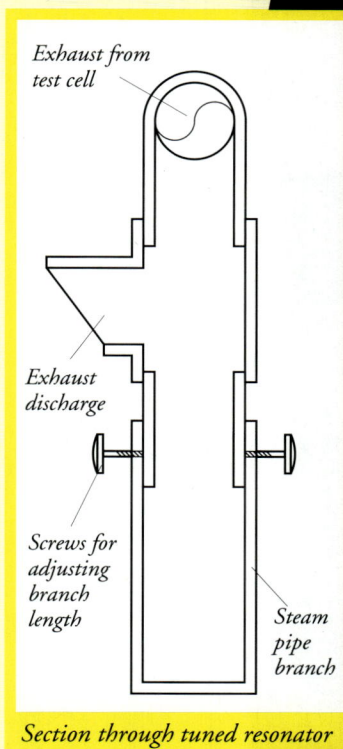

Exhaust from test cell

Exhaust discharge

Screws for adjusting branch length

Steam pipe branch

Section through tuned resonator

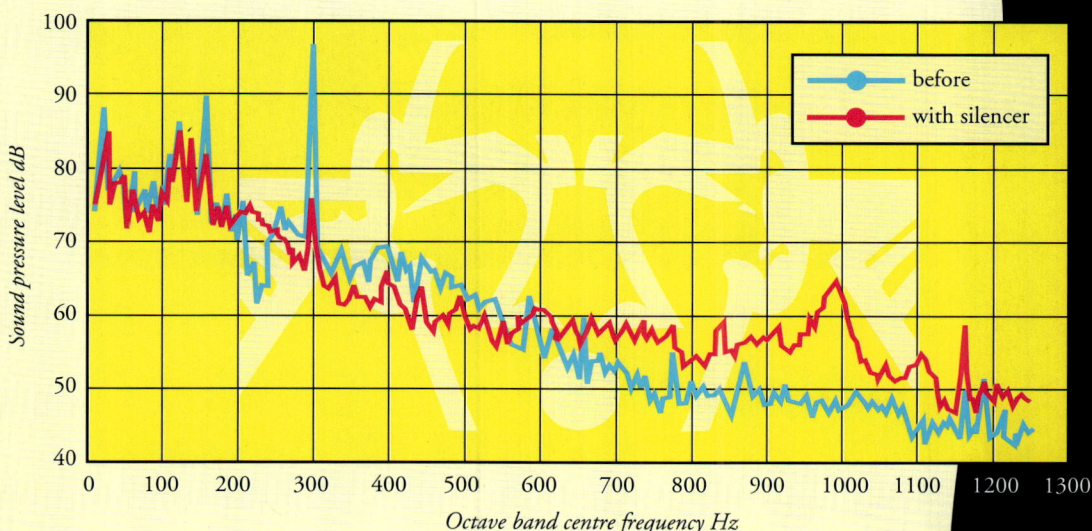

Octave band centre frequency Hz

The result

The trace shows the frequency analysis of the exhaust noise before and after the addition of the silencer, measured close to the exhaust.

Information supplied by
Coventry City Council
and **Brico Engineering Limited**

24
Reducing noise in a dump truck

The problem

The cabs of large dump trucks operating on mineral working and construction sites were found to have noise levels of about 95 dB(A).

The cabs were situated immediately beside the engine compartments and separated from them by plain sheet steel bulkheads. The vehicle construction itself was one of an all-welded, undamped steel plate that resulted in noise and mechanical vibration emissions from the engine and gearbox.

The solution

A three-fold approach to noise reduction was taken:

- the large surface areas of the cab, roof, bulkhead and door panels were treated by the application of 10 mm thick bitumastic damping pads to reduce resonant vibration;
- a proprietary sound barrier mat consisting of a lead/fibreglass laminate was applied to the floor and engine bulkhead;
- an acoustically absorptive foam was applied to all available surfaces inside the cab to reduce reverberant sound build-up.

Damping sheet and sound barrier mat on engine bulkhead

Damping sheet on roof and doors

Acoustically absorptive treatment to all cab interior

Sound barrier mat on floor panels

Dump truck

The cost

About £1500 (a new vehicle can cost over £200 000).

The result

A noise reduction of up to 11 dB(A).

Information supplied by
EMD Services

The problem

A company using tracked drag-lines in quarrying found that drivers were being exposed to noise levels of 96 to 100 dB(A).

The particular problem was that the main drive engine, in this case a 220 hp diesel, was located in the same all steel 'box', separated from the driver's cab only by a sheet steel bulkhead containing a door. Engine noise was being transmitted through the cab bulkheads and floor panelling. Mechanical vibration from the engine and drag-line drives also travelled easily through the steel structure to add to noise levels in the driver's compartment.

The solution

The first stage in the noise reduction programme was to reduce the mechanical vibration. This was achieved by applying pads of a proprietary damping compound to all available sheet steel surfaces in the engine compartment and driver's cab.

Noise transmissions through the cab bulkhead and floor were then reduced by applying a lead/fibreglass laminate to both faces of the bulkhead and to the underside of the cab floor.

Finally, reverberation in the driver's cab was reduced by applying PVC-faced acoustic foam to the roof panels and other available internal surfaces.

Drag-line cab

The cost

About £1500.

The result

A noise reduction of up to 10 dB(A). The treatment can be applied to similar site plant such as shovels, bulldozers and scrapers.

Information supplied by
EMD Services

26
Reducing noise in crew cab road vehicles

The problem

Noise levels in crew-carrying road vehicles can be high enough to cause concern. In some untrimmed vehicles such as refuse collectors, levels have been measured as high as 103 to 105 dB(A). One local authority operating a fleet of such vehicles commissioned a noise reduction exercise to determine whether it would be reasonably practicable to effect a significant reduction without compromising the vehicle's road worthiness.

The primary source of noise came from the engine. However, the cabin was also subjected to noise from the transmission and hydraulic power pack coming through the relatively lightweight sheet steel floor and toe plates and the rear bulkhead.

The solution

This secondary noise was reduced by securing a lead/fibreglass laminate sheet to the underside of all floor, toe and bulkhead surfaces. Reverberation was also reduced by lining the roof and all available internal surfaces of the doors and the front and rear bulkheads with an expanded acoustic polyurethane foam faced with a washable porous PVC film.

Internal views of truck with absorptive lining

The cost

About £1300.

The result

A noise reduction of 4 to 9 dB(A). Similar reductions have since been achieved on other types of crewed vehicles such as fire and maintenance tenders.

Information supplied by
EMD Services

The problem

The hydraulic power pack of a 250 tonne billet hydraulic guillotine was found to be generating noise levels of between 90 and 95 dB(A). Noise was generated through vibration transmission from the motor pump unit to the machine frame (and through rigid pipework to the remainder of the machine) and impacts from the valve bank. An acoustic enclosure was unsuitable because its effectiveness would have been limited by the transmission of vibration through the pipework.

The solution

It was decided to reduce the noise by isolating the noise sources from the guillotine. This solution was considerably cheaper than an enclosure and did not affect access.

The motor and pump units were mounted on a rigid frame, isolated from the machine. The rigid pipework was interspersed with flexible couplings. The valve bank was isolated from the machine frame using a cork-rubber composite.

Hydraulic guillotine after isolation

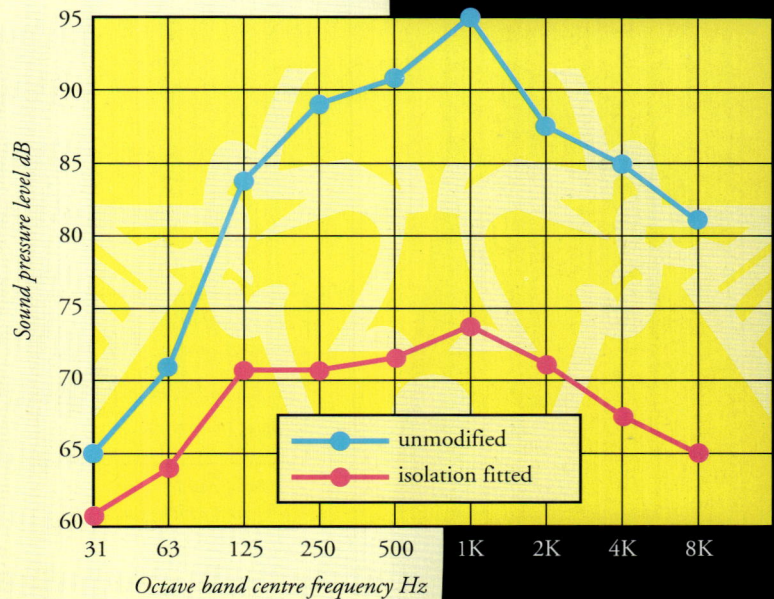

The cost

£1500.

The result

A noise reduction of at least 17 dB(A).

Photograph courtesy of
Joseph Rhodes Limited

Acoustic consultants were
Industrial Noise and Vibration Centre

28
Pneumatic impact press noise reduction

The problem

A bench-mounted pneumatic impact press used for riveting small switch components was found to be generating excessive noise levels. The main peak noise emissions were from the release of compressed air at the actuator exhaust and from the impact of the metal actuator ram as it struck the metal tool ram.

The solution

The compressed air exhaust noise was reduced after a proprietary lightweight silencer had been fitted. The flow was passed through a porous polythene cap of the type indicated in the diagram below.

The impact noise was reduced by cushioning the contact between metal surfaces with an insert of damping compound, again as indicated in the diagram. The most effective acoustic performance was by an insert of 8 mm thick nitride rubber. However, this showed early signs of wear and some loss of force transmission to the workpiece. The best compromise between impact force transmissions, wear and noise reduction was obtained by 8 mm thick Betathene, a proprietary urethane elastomer.

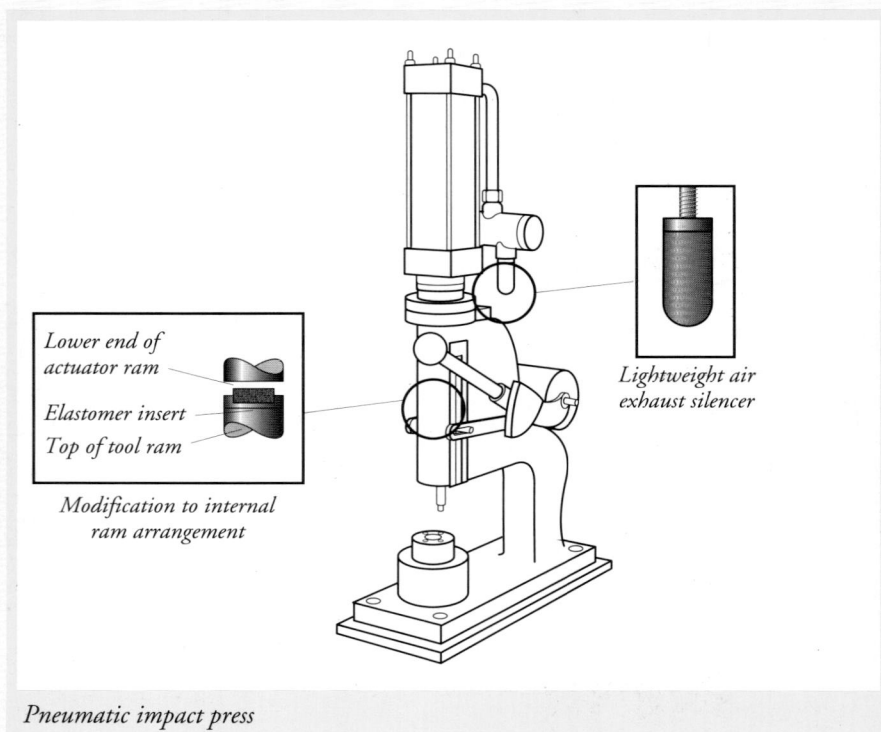

Lower end of
actuator ram

Elastomer insert

Top of tool ram

*Modification to internal
ram arrangement*

Lightweight air
exhaust silencer

Pneumatic impact press

The cost

About £50.

The result

The compressed air exhaust silencer reduced peak noise levels at each operating position by some 7 dB(A), with a corresponding reduction of some 5 dB(A) over about five cycles of the machine during continuous operations.

The cushioning of the impact noise gave a further reduction in peak noise levels of 5 dB(A) and about 4 dB(A) while the machine was running continuously.

Overall, there was a 9 dB(A) reduction from the two modifications, achieved with no significant effect on the overall working efficiency of the machine.

Information supplied by
Lucas Industries Noise Centre

The problem

One drawback of using bench-top, pneumatic impact presses is the high noise resulting from the metal actuator striking the tool rams to transmit sufficient force to the workpiece.

The solution

One way to avoid this is to use a squeeze press. With this, a pneumatic actuator brings the ram onto the workpiece and then applies the force required to perform the operation through a second steady pressure, so avoiding any unwanted impact.

The squeeze press illustrated below can exert a forming load of up to 2 tonnes. The pneumatic actuator assembly can be fitted retrospectively to certain types of impact presses. It includes a safety feature preventing the full secondary working load from being applied until the ram actually touches the workpiece.

Noise generated by compressed air exhaust flow can be dealt with by fitting a standard proprietary air exhaust silencer.

Pneumatic squeeze press

The cost

About £1400 for a new press, and £500 to retrofit an existing press.

The result

20 dB(A) has been claimed compared with levels from similarly sized impact presses.

Information supplied by
Fort Vale Engineering Limited

30
Machining alternator castings

The problem

Machining an alternator end casting was producing noise levels of 104 dB(A), principally in a high pitched 'squeal' radiating from the casting. Overlapping hanging strips had been fitted around each machine to form a partial acoustic screen. However, these had been cut away at the operator position to improve access.

The noise arose from resonant vibration of the casting, excited by cutting forces, with evidence of chatter marks on some of the machine surfaces. The mounted casting exhibited very low damping characteristics during machining.

Casting and bungee

The solution

The solution was to apply a simple damping treatment to reduce the casting's vibration.

To test the method, an inexpensive rubber bungee was wrapped around a casting during machining. Subsequently a pair of damping straps were designed to provide a more durable solution. These were easily fitted to the casting by the operator while the preceding casting was being machined.

The cost

£40.

The result

An overall noise reduction of 16 dB(A) with the 'squealing' component being reduced by 30 dB. In addition, the quality of the cut improved and the machining time for each cycle was reduced.

Photograph courtesy of
Industrial Noise and Vibration Centre

Consultants were
Industrial Noise and Vibration Centre

Dump truck lined with abrasion-resistant rubber

The problem

During quarrying, noise levels of 96 dB(A) can be produced when the shovel or conveyor drops its first loads of material onto the bare metal of the body of dump trucks.

The solution

One company carrying out large-scale quarrying operations reduced noise levels by lining the bodies of its 20-tonne dump trucks with abrasion-resistant rubber. Sheets 50 mm thick were used, and moulded so that a profiled section of 200 mm overall depth was obtained. The sheets were then attached mechanically to the body of the truck.

Trucks were loaded by emptying a series of large shovels, each carrying up to 5 tonnes of material. The force of the impact, previously borne by the bearings, was now absorbed by the lining.

The cost

£26 000 per truck. In an industry where each vehicle can cost over £250 000, any extension in service life is valuable.

The result

The lining reduced noise levels by 14 dB(A). The treatment also significantly increased truck body life.

Information supplied by
Trellex Limited

32
Material change - block making machine

The problem

In a firm manufacturing paving blocks, the blocks were moved along the line on wooden pallets. The pallets were propelled by a series of metal catches underneath the machine frame, which rose up to engage the pallet and to push it forward to the next stage of the process. At the end of the stage, the catches fell to a horizontal position and slid back along the line to the next pallet. They then rose to engage the new pallet to repeat the process.

Noise levels of over 100 dB(A) were generated when the catches snapped back to the horizontal and fell onto the metal bar.

The solution

The solution was to replace the metal catches with similarly shaped components moulded from a tough polyurethane polymer. The new catches also proved to be almost as durable as the original steel components.

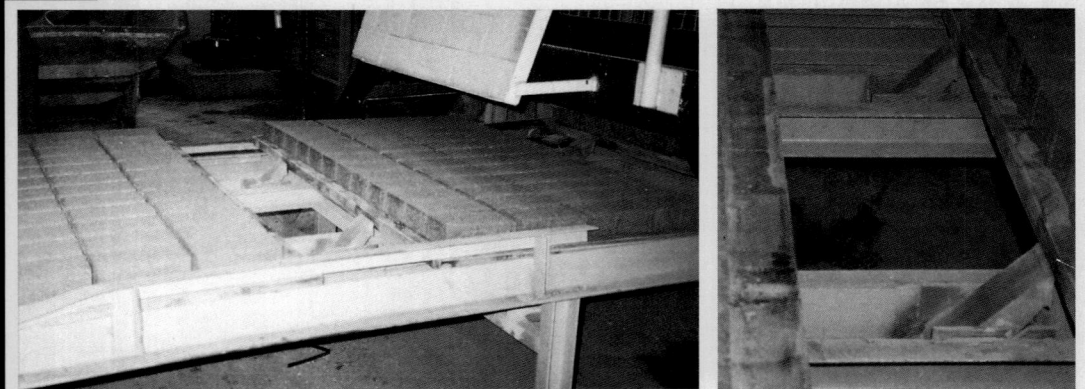

Paving block machine

The cost

About £10 to £20 per catch.

The result

Noise levels were reduced by about 5 dB(A). Impactive peaks were effectively removed.

Photographs courtesy of
Redland Precast

Equipment supplied by
Hallam Polymer Engineering Limited

The problem

Many process machines radiate noise due to multiple impacts from the components being handled. In some cases noise levels can be reduced by lining hoppers, chutes and conveyors, but in others the ambient conditions do not allow such techniques. Alternative approaches then have to be considered.

One such example was a machine used in processing cocoa beans. The beans were transported along a vibrating tray beneath an array of gas burners. Impacts between the hard bean cases and the steel conveying tray produced noise levels of between 88 and 90 dB(A). Resonance in the cavity between the vibrating bean tray and the bed of the machine was also a contributory factor.

The conventional remedy of cushioning impacts from the beans by providing a lining of rubber or similar material for the bean tray could not be used here because of the high temperatures.

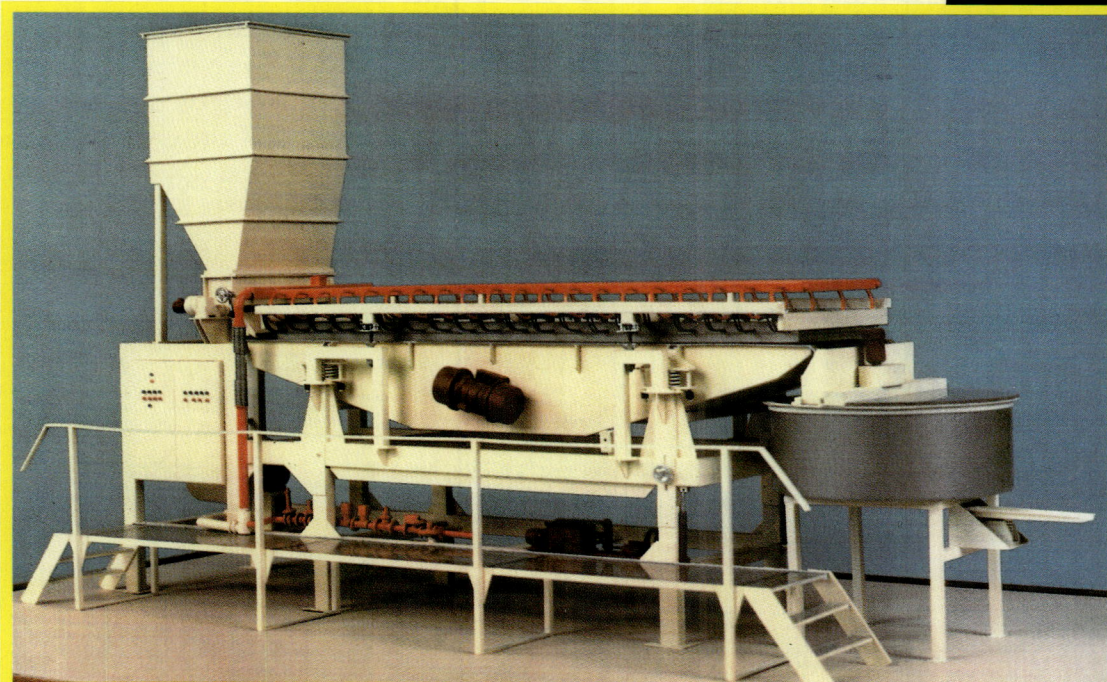

Cocoa bean processor with damped tray

The solution

It was decided that the most effective way of reducing the amount of vibratory energy reaching the external radiating surface of the machine would be to reduce the response of the structure at the point of impact, ie at the bean tray itself.

This was achieved by applying point-fixed damping to the tray in the form of 2 mm thick steel plate, plug-welded to the underside of the bean tray between its existing tubular steel stiffeners. Acoustic resonance in the void beneath the tray was reduced by filling it with 96 to 128 kg/m^3 mineral wool slab.

The cost

About £2000.

The result

A noise reduction of 10 dB(A).

Information supplied by
Sound Research Laboratories Limited, Sudbury

Photograph courtesy of
Micronizing Company (UK) Limited

34
Very high performance drywalling

The problem

Many production areas create high noise levels. To reduce risks to workers' hearing, it is often desirable to have acoustic separation between the areas, for example to:

- separate large process fans from workers;
- provide sound havens;
- allow noise testing;
- concentrate high noise fabrication processes into one area.

The solution

A very high performance drywall partition has been developed to give considerably higher sound transmission loss than a traditionally rendered cavity brickwork wall but with less than one-third of the superficial weight.

The drywalling is made of two independent wall structures with metal studs 600 mm apart. One wall has two layers of 15 mm plasterboard on the outside and a single layer of 19 mm plasterboard on the inside. The cavity is partially filled with 100 mm of dense mineral wool insulation.

The secondary skin of the drywall is made up of two layers of 15 mm plasterboard on the outside of the studding and a further 100 mm of dense mineral wool insulation on the inside. The width of the cavity and the absorption by the mineral wool insulation within it provide good attenuation of transmitted sound energy.

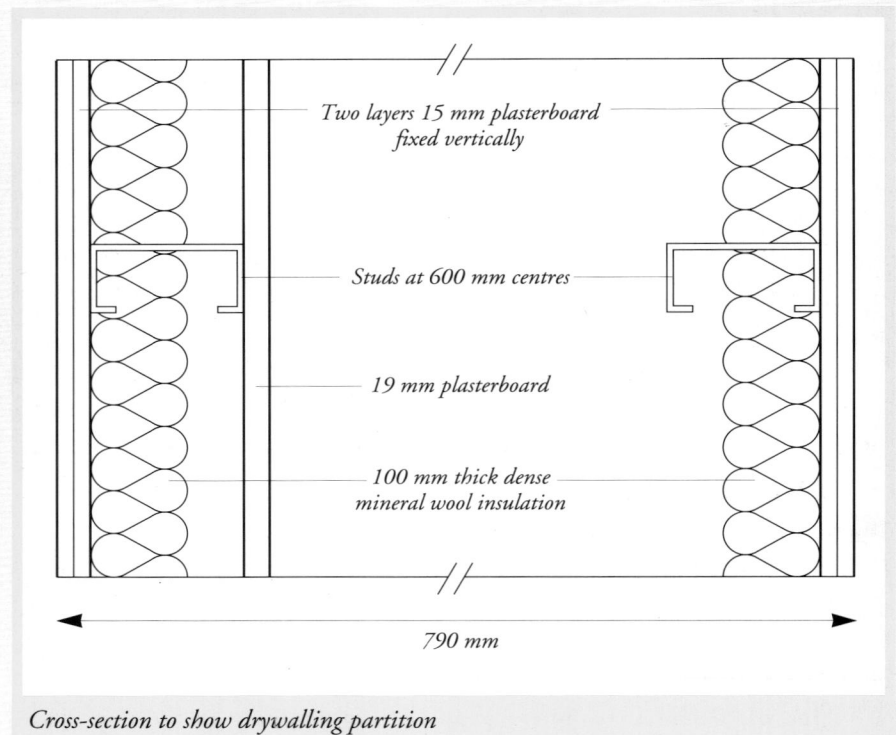

Two layers 15 mm plasterboard fixed vertically

Studs at 600 mm centres

19 mm plasterboard

100 mm thick dense mineral wool insulation

790 mm

Cross-section to show drywalling partition

The cost

Approximately 50% of the cost of conventional partitions.

The result

An average sound reduction index of 77 dB at one cinema complex. The partition is claimed to be equally suitable for industrial areas.

Designed by
**Vernon Cole Associates
(Acoustic Consultants)**

The problem

An engineering firm, manufacturing components for conveying systems, has developed a batch collation system which takes in nails from an adjacent washer and aligns them using a vibratory feed system. The nails are first shaken up a spiral to the top of the machine and then aligned into collation boxes along a vibrating chute.

Although very effective, the machinery generated noise levels of over 100 dB(A) mainly from the impact of the nails against the steel spiral and collation boxes.

The solution

The solution was to enclose the entire feed system. Another idea - to use alternative materials for the collation boxes, and so remove the metal-on-metal impact - was rejected because it would have led to unacceptable wear and the loss of accuracy in the batching process.

The enclosure was manufactured from standard acoustic panels consisting of 30 mm and 50 mm thick mineral wool between a plain and a perforated steel sheet.

The product entered the enclosure along a chute that enabled nails to be poured from the adjacent washing machine into the feed system. The design included ample provision for access, including an ingenious arrangement on the corner where either portion of the door could be opened as required.

The main access was above the collation boxes where glazed panels were placed on hinges with handles for lifting clear. Other features of the enclosure included a ventilation fan on the roof to prevent overheating and the chamfering of one corner to allow operators to get closer to the equipment.

Enclosed nail sorting machine

The cost

About £2500.

The result

A noise reduction of 26 dB(A).

Photographs courtesy of
Mato Industries Limited

Equipment designed and manufactured
Euro Acoustics Limited

36
Reducing combustion noise by flow control

The problem

Typical noise levels generated by a dryer oven, used in malting grain, reached up to 95 dB(A). A low-frequency 'boom' at a tone of 82 Hz dominated. As the boom occurred only at low firing, it was diagnosed as flame instability that developed into an oscillation at a resonant frequency of the combustion tube. The precise cause of the instability was determined to be an excessively weak ratio of fuel to air at low firing.

The solution

To overcome this problem, a motor-controlled valve, previously used for controlling weld fume extraction, was adapted and fitted to the combustion fan intake. Triggered by a micro-switch, the valve was used to reduce the air flow at low loads and so improve the combustion characteristics.

The cost

£100 per unit.

The result

An overall noise reduction of about 8 dB(A) with the low-frequency tone being reduced by 21 dB. The bought-in units had no effect on operations, could be retrofitted and satisfied hygiene requirements.

Dryer oven burner

Consultants were
Industrial Noise and Vibration Centre

The problem

Shoe manufacture is a labour intensive process, involving operators in semi-automatic tasks during most production stages. Much of the machinery is based on well-established, efficient technology. However, it is often likely to generate high noise levels.

The solution

A major shoe manufacturer has designed a new paring machine, incorporating the following features:

- lining the outer casing of the machine with 50 mm thick, absorbent foam to increase sound reduction across the panels and to dampen mechanical vibration of the outer skin;
- installing the cutting head system on anti-vibration mounts to prevent vibration transmission from the blade and shoe to the outer casing;
- placing the dust extraction system on the outside of the machine. This can be positioned away from the operator.

The cost

About £150 per machine.

The result

An early prototype provided a reduction of up to 5 dB(A). It is hoped that subsequent design changes will further reduce noise levels.

Operative using new paring machine

Photograph courtesy of
Clarks International

38
Reducing foundry pipe spinning noise

The problem

An automatic pipe spinning machine in a foundry was generating noise levels in excess of 90 dB(A). The pipe mould was housed in a sheet steel cabinet with the pipe being spun at high speed before molten metal was fed into the centre. The spinning action forced the molten metal to the outside of the mould while the cabinet was filled with water to cool the mould and solidify the spun pipe.

The primary source of noise was mechanical vibration in the hydraulic drive mechanism and main mould bearings being radiated from the sheet steel surfaces of the water cabinet and the machine support structure. Noise also came from the cabinet roof, steam vents and the hydraulic power pack.

The solution

A number of modifications to the machine were made, including the following:

- stiffening the main bearing supports;
- isolating the supports from the main machine chassis, and the chassis from its supporting beams using proprietary machine mounting pads;
- sand-filling the chassis support beams and isolating them from the floor with similar pads;
- isolating the water cabinet from the machine support structure using the machine mounting pads;
- using rubber-bushed hinges on the water cabinet roof panel;
- fitting duct attenuators over steam vent outlets in the water cabinet roof;
- relocating the hydraulic power pack outside the shop.

Pipe spinning machine after modification

The cost

About £2000.

The result

An overall noise reduction in the mould shop of 6 to 7 dB(A).

Photograph courtesy of
Glynwed Foundries

Consultants' report supplied by
Sound Research Laboratories Limited, Wilmslow

The problem

Strand pelletisers are used in the plastics industry to convert continuous strands of plastic into small pellets for use in subsequent manufacturing processes. The most important part of a pelletiser is its cutting head where the plastic strands are fed by rollers into the path of a high-speed, multi-bladed rotating head.

Noise levels from a pelletiser can exceed 90 dB(A). Noise is generated by the impact of each blade against the strands and the alternate compression and expansion of air as the moving knives pass the fixed bed knife edge.

The solution

A company manufacturing strand pelletisers has developed a series of new machines which incorporate a number of important noise reducing features.

On examining the cutting operation, it was revealed that the speed of rotation could be reduced by increasing the number of blades fixed into the cutting head. It was found that a helical blade, which would pass progressively across the bed knife edge cutting one strand at a time rather than all simultaneously, would further reduce noise levels. Additional benefits of the helical cutter were found to be reduced wear and a reduced need to sharpen blades.

The machine casing was also redesigned, reducing both mechanical and airborne noise radiation. Anti-vibration treatment was adopted, with the cutting head installed on a base isolated from the rest of the machine.

The cost

About £25 000 for a new pelletiser and about £11 000 to upgrade a conventional pelletiser.

The result

A noise reduction of at least 10 dB(A).

Views of redesigned strand pelletisers

Photographs courtesy of
John Brown Plastics Machinery Limited

Equipment designed and manufactured by
John Brown Plastics Machinery Limited

The problem

A commonly encountered noise source in a process plant is the high-pressure, low-volume centrifugal fan units used typically to supply air for product conveying systems. The high-pressure airflow in these fans can generate high levels of noise, usually with strong tonal content.

A major chemical manufacturing company anticipated noise levels above 95 dB(A) from a pneumatic conveying fan to be installed in the centre of an occupied production floor. It opted to control the noise by installing an acoustic enclosure. Other noise control approaches, eg modifying the fan design or relocating the unit, were considered impracticable. Enclosure, however, did pose its own design problems. The design had to:

- allow immediate access to the fan drive motor, belts and pulleys;
- provide adequate airflow through the enclosure for cooling the high-speed drive motor;
- allow the enclosure to be dismantled completely when the fan or motor needed to be replaced.

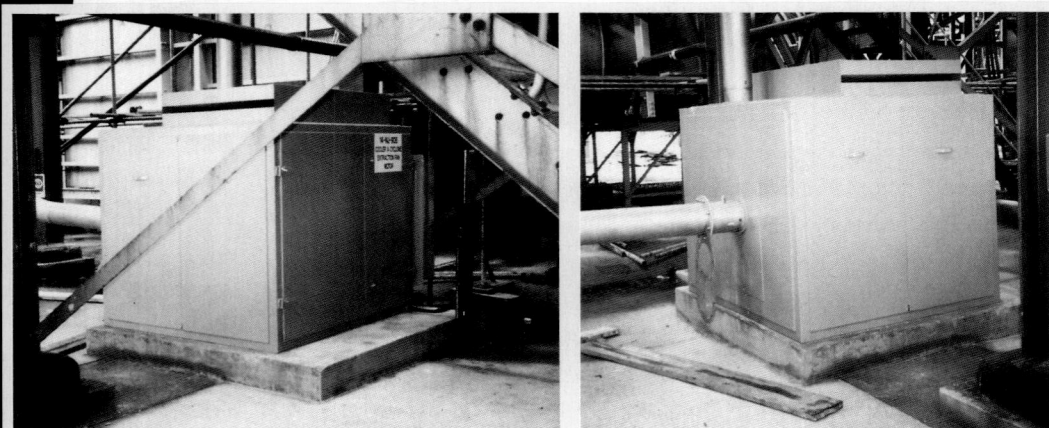

Enclosed pneumatic fan

The solution

The enclosure was built up of 50 mm thick acoustic panels consisting of a plain steel sheet outer skin, a perforated steel sheet inner skin and an interleaving acoustically absorptive lining of semi-rigid mineral wool slabs. The panels were mounted on a steel frame with hinges and quick release toggle catches for access to the fan drive, with bolted fixings where less frequent access was required.

Cooling air was passed into and from the enclosure via two slots in the acoustic panelling, each covered with an acoustically-lined duct. One of the slots was positioned close to the internal motor cooling fan so that it could draw air into the enclosure.

The cost

About £3000.

The result

A reduction in noise levels of about 20 dB(A).

Photographs courtesy of
ICI Chemicals plc

The problem

Pneumatically-driven screwdrivers can produce high noise levels of over 80 dB(A), particularly when used in large numbers and in confined spaces, eg on production lines. In general, there are two noise generating mechanisms:

- exhaust air sound caused both by air flow variation as the air flows past the motor and by aerodynamic sound generation in the exhaust-air channel;
- vibration-radiated sound from the surface of the machine produced by the moving parts of the machine and by the flow of air inside the machine.

The solution

To reduce the noise generated by exhaust air, a manufacturer of pneumatic tools modified its range of screwdrivers to incorporate a hose to pipe the air away from users. This also removed any oil fog carried in the exhaust air.

As the exhaust noise was reduced, the vibration-induced noise radiation became more significant. To overcome the problem, the company investigated the precise mechanism of vibration driving, which resulted in a major change of screwdriver design.

The rear section of the tool was screwed into an outer housing of anodised aluminium. An inner housing containing the air motor and gears was carried on axial and radial rubber elements to isolate it from the outer casing. The motor rested on the axially positioned elements but the radial O-rings prevented direct metallic contact between casing and motor.

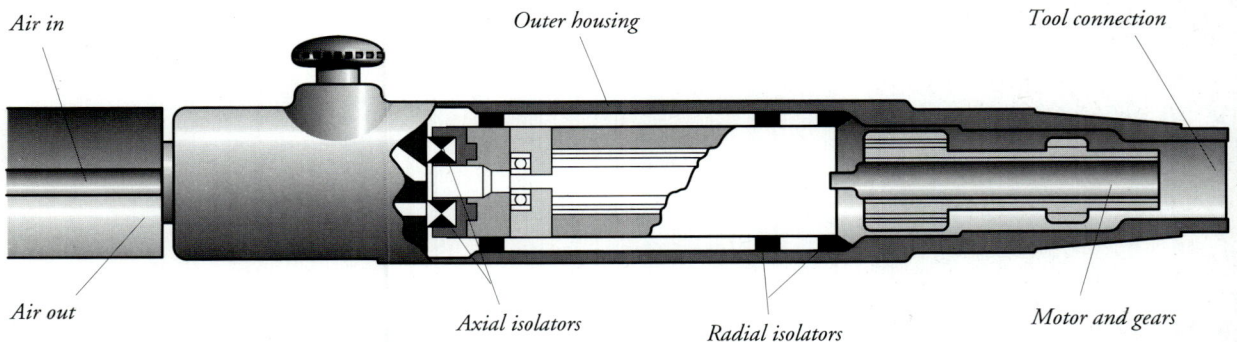

Cross-section of a screwdriver motor isolation

The cost

Negligible difference over traditional screwdrivers.

The result

In a controlled test on an individual machine, the treatment resulted in noise levels being reduced by up to 7 dB(A).

Photograph courtesy of
Atlas Copco Tools Limited

Equipment designed and manufactured
Atlas Copco Tools Limited

The problem

Pneumatic tools can emit high noise levels, much of it from the exhaust of the compressed air that drives the motor. Each time one of the motor vanes passes an exhaust port, the air contained between the vane and the next is emptied into the exhaust system. The rotation of the vanes creates a series of pressure waves moving through the exhaust pipe. Additionally, there is usually a degree of aerodynamically generated noise caused by the velocity of air passing through a narrow exhaust hole.

While conventional reactive silencers are effective at reducing noise emission, they are not generally used because the back pressure they impose on the air flow adversely affects the power output and air consumption.

In one type of pneumatically-driven tool, a hand-held vertical grinder, noise levels of 85 dB(A) can be reached during machine idling, with peak values of over 90 dB(A) during start up or shut down.

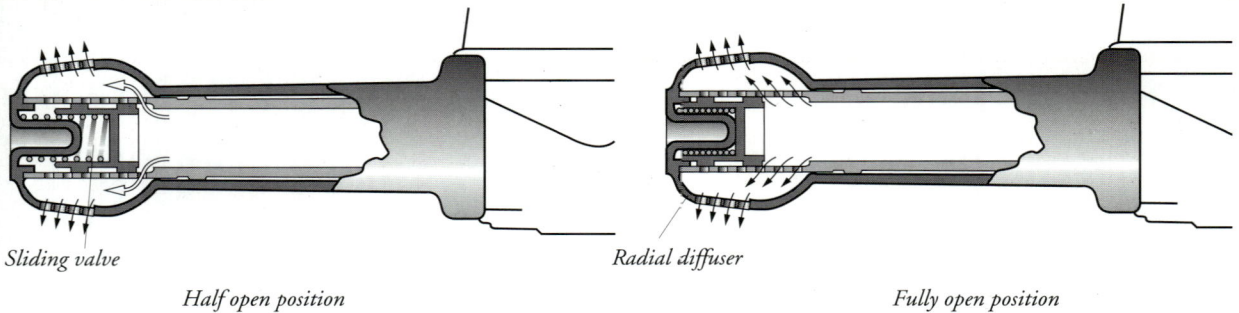

Sliding valve

Half open position

Radial diffuser

Fully open position

Cross-section of a grinder exhaust active valve

The solution

One manufacturer of such tools has achieved good attenuation without causing excessive back pressure by using a spring-loaded valve in the exhaust. As the pressure of the exhaust air varies, the valve opens or closes the exhaust opening in quick response, as shown below. The mechanism keeps the back pressure in the motor exhaust roughly constant and virtually independent of the air flow.

The high-frequency noise generated when air passes through the radial holes is attenuated by means of a diffuser at the end of the handle. The pattern of holes in the diffuser is designed to generate a very small pressure drop and so the air speed in the outlet hole is slow.

The solution also means that the valve automatically closes when the machine is not in use. This protects the machine from contaminants which could find their way in through the exhaust.

The cost

Negligible difference over traditional grinders.

The result

A noise reduction of between 5 and 10 dB(A).

Grinder

Photograph courtesy of
Atlas Copco Tools Limited

Equipment designed and manufactured by
Atlas Copco Tools Limited

The problem

A strip-fed, 115 tonne power press, installed to stamp oil filter base plates, generated noise levels of 99 dB(A), principally in 'bell-like' tones radiated by the flywheel.

The solution

The vibration response of the flywheel was checked to confirm that its natural frequencies corresponded to the dominant tones in the radiated noise. A small electrodynamic vibrator was used to excite each resonant frequency of the flywheel to determine how the system worked and where the maximum vibration could be found.

The level of acoustic radiation was determined from the mechanical vibration in the flywheel. It was concluded that this could be reduced by using dynamic vibration absorbers fixed to the flywheel where the vibration was at its highest. The absorbers consisted of mass on a spring, tuned to a specific frequency. When attached to the flywheel, it transferred its vibratory motion into vibration of the small masses.

In this case, small mild steel discs and plates, mounted on rubber-bonded cork, were used as absorbers. They were bolted on to the flywheel at positions of maximum vibration and then tuned to the resonant frequencies. The flywheel itself needed no modification.

The cost

£15 per absorber.

The result

An overall noise reduction of 10 dB(A) and virtual elimination of the flywheel tones.

Flywheel

43
Dynamic absorption of power press vibration

Consultants were
Industrial Noise and Vibration Centre

44
Shielding a printing press

The problem
The dominant noise source from a printing press came from the second print head air drum. Secondary noise sources included drying fans and suction systems. Together these generated noise levels of up to 95 dB(A). Given the size of the press (some 3 m x 8 m) and the access requirements, an acoustic enclosure was ruled out.

The solution
As the majority of energy was high frequency, and therefore directional, the solution was to shield the major noise sources with acoustically absorptive material, strategically positioned in the direction of maximum noise radiation.

Plain polycarbonate sheet was used for the viewing panels and 20 mm acoustic foam was placed inside the existing guards to provide absorption of the radiated noise. The cooling air passages to and from the press were silenced by adding cowls lined with absorbent material.

Shielded printing press

The cost
Up to £1000.

The result
A noise reduction of about 11 dB(A).

Photograph courtesy of
Medica Packaging Limited

Noise survey conducted by
Industrial Noise and Vibration Centre

The problem

While considering the implementation of a noise control programme, a major soft drinks manufacturer was advised that a pump used for transferring fruit juice concentrate from 200 litre drums to the product mixing tanks was producing noise levels of 103 dB(A).

The solution

A preliminary inspection showed mechanical deterioration of the pump which required a complete overhaul to include:

- rewinding the drive motor;
- replacing the motor bearings;
- checking and cleaning the pump bearings;
- repacking the pump glands.

The cost

Nominal costs for staff time.

Pump

The result

A noise reduction of 27 dB(A).

Reducing pump noise by good maintenance

Report details supplied by
Sound Research Laboratories Limited, Wilmslow

46
Use of absorption in a noise control programme

The problem

Noise levels from high-speed presses operating in the production area of a health care company manufacturing metered dose aerosols were generally over 85 dB(A). Most of the operators' exposure was due to direct radiation from the nearest presses, but reflected noise also built up as reverberant sound energy.

The solution

As part of their overall noise control programme, the company decided to install a network of absorbers suspended from the roof of the building. The absorbers had to be as low as possible for maximum acoustic effect but could not interfere with the fire alarm, power and sprinkler systems already placed in the ceiling.

Approximately 1400 absorbers were suspended from the eaves. Each unit was 900 mm long x 600 mm wide x 50 mm thick with a perforated steel case and plastic wrapping to prevent oil getting into the absorbent material.

Hanging absorbers

The cost

About £30 000

The result

After installation, the reverberation time of the press shop fell and overall there was a noise reduction of 4 dB(A).

Photographs courtesy of
3M Neotechnic Limited

Equipment supplied and installed by
Ash Peace Limited

The problem

Conventional steel acoustic enclosures can reduce noise significantly but can sometimes restrict access. Careful design of the doors and how the product enters the enclosure can often overcome this. However, it can be difficult to define the exact access requirements for some machinery.

A company producing air-conditioning systems had two high-speed automatic presses generating noise levels of up to 93 dB(A). The company wished to reduce the noise radiated to adjacent areas but needed to maintain all-round access.

The solution

The solution was to install a PVC strip enclosure around each machine. This consisted of a steel framework, approximately 7 m long x 7 m wide x 3 m high, from which 4 mm thick strips of clear PVC plastic were hung. The roof of the enclosure was made of a 75 mm thick acoustic tiled ceiling laid into a self-supporting grid system, and overlaid with PVC sheet.

Each strip was 400 mm wide with the fixing arranged so that each one overlapped its neighbours by 50%, effectively doubling the thickness of the enclosure wall. The strips were secured at the top into double aluminium tracks allowing sections to slide open, enabling product and personnel access at any point. The transparency of the enclosure was also considered to be an important safety feature by the company.

The cost

About £5000

PVC strip enclosures in the production area

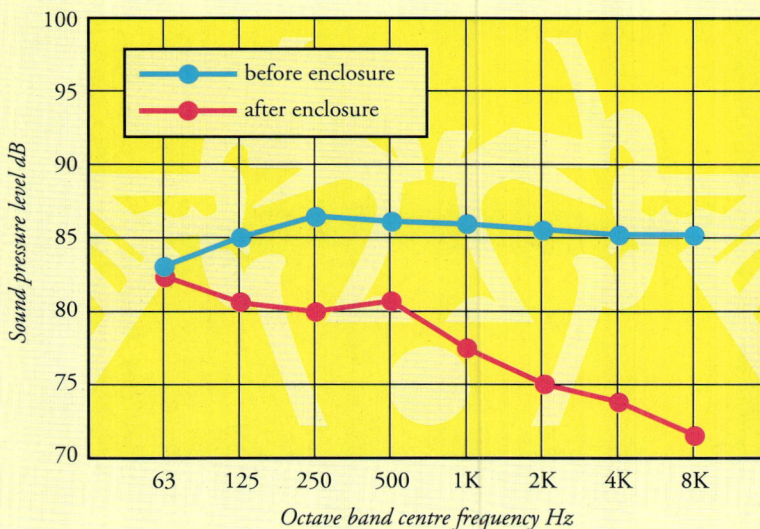

Sound pressure level dB

before enclosure
after enclosure

Octave band centre frequency Hz

The result

A noise reduction of about 10 dB(A).

Photograph courtesy of
Dunham-Bush Limited

Enclosure supplied by
**Mainline Industrial Products
Limited**

48
High speed press noise - modifying safety guarding

The problem

A health care company manufacturing medical containers used a process of multi-stage forming on high-speed, automatic, coil-feed transfer presses. The presses consisted of a number of cam-driven, mechanical hammers mounted on heavy steel base frames, factory fitted with enclosures acting both as a safety guard and to reduce some noise emission.

It was impossible to change the noise generating mechanism of the press, nor would increasing the acoustic absorption have resulted in sufficient noise reduction. In addition, the shop ran on 24-hour shifts, preventing any prolonged acoustic examination of a given press running in isolation.

The solution

A sound intensity survey of each press was carried out, enabling the acoustic energy output of any selected part of the machine to be measured in isolation.

The survey identified those presses that were the major contributors to overall shop noise and how that output was distributed over the various panels of the machine. The noise sources and resulting treatment were as follows:

- the press base-frame was clad with preformed acoustic panels;
- acoustic covers were placed over the slots provided for raising the access doors on the roof;
- lightweight polymer glazing on the access doors was replaced by double-skin polycarbonate;
- the enclosure was isolated from its base by pads of anti-vibration material;
- hinged acoustic enclosures were fitted over the coil feed slot.

Modified safety guarding around a press

The cost

About £2000 per press.

The result

A noise reduction of up to 11 dB(A) per press.

Photograph courtesy of
3M Neotechnic Limited

Noise control survey conducted by
Ian Sharland Limited

The problem

High-speed, strip-fed punch presses are a common source of noise. In one company, a press used for stamping electrical trip-catch components caused levels of 101 dB(A) when running at 271 strokes per minute.

The major noise radiating areas were the legs on which the press frame sat. As these were welded, they were prone to transmit vibration with little or no damping.

The solution

To reduce the transmission of mechanical vibration from the press frame to its supporting legs, it was decided to insert 6 mm thick composite pads between them. The area and arrangement of the isolating material was selected to provide a resonant frequency of 65 Hz on both legs, below both the frequency of all major structural resonance and the frequency of maximum acoustic radiation. Care was taken to ensure that there was no mechanical link between the press frame and legs.

Other work was carried out to reduce noise radiated by the press manufacturer's acoustic guards which had been subjected to the same vibration. This was achieved by applying a layer of self-adhesive damping sheet to the sheet metal surfaces of the guards.

The cost

£50 for materials.

The result

A noise reduction of about 9 dB(A). The treatment had no effect on machine operation, access or maintenance.

High-speed press

49
Anti-vibration treatment of high-speed presses

Consultants were
Industrial Noise and Vibration Centre

The problem

Five overhead reciprocating presses with ratings ranging from 40 to 120 tonnes were located within a small batch production press shop measuring 15 m long x 10 m wide x 8 m high. The internal surfaces were acoustically reflective, with a concrete floor, blockwork walls and a steel clad roof. The noise level was found to be in excess of 90 dB(A).

The presses were used to produce quite small batch runs and so had an unusually high down time. As a result, an operator's overall noise exposure was strongly influenced by the noise from neighbouring presses.

It was inappropriate to enclose the presses because of the need to feed strips manually across the tool and to change the workpiece. The reflective finishes within the shop resulted in a low rate (approximately 3 dB) of reverberant fall off between one operating press and an idle press on the other side of the shop.

Reciprocating press with acoustically-clad wall behind

The solution

The presses were positioned very close to the wall. Reflection of sound from the wall acted as a major contributor to the reverberant sound pressure level. As a result, the absorptive treatment was concentrated on the walls. An acoustic lining consisting of 50 mm thick 45 kg/m³ mineral wool, retained behind a galvanised perforated steel sheet, was fixed to the walls between 1 and 4 m above the floor. The mineral wool was wrapped in plastic film to keep oil mist out.

Three mobile acoustic screens, 2 m wide and 2.4 m high, were also used to reduce direct line-of-sight radiation between presses. The screens were faced with mineral wool retained behind perforated sheet steel to prevent the reflection of sound energy.

The cost

£8000 for the absorptive lining and £2000 for the mobile screens.

The result

The acoustic cladding to the walls increased the reverberant fall off across the press shop from 3 to up to 11 dB(A) and the mobile screen gave approximately 3 dB(A) additional protection.

Photograph courtesy of
John Crane UK Limited

Equipment manufactured by
Ian Sharland Limited

Acoustic consultants were
Ajax Health and Safety Services

The problem

Rotary blowers often produce high levels of sound at discrete frequencies, related to the rotational speed of the unit. If these discrete frequencies are at the low end of the spectrum they may be difficult to attenuate.

Traditional absorptive silencers work well at middle and high frequencies but their performance drops off below 250 Hz. Reactive silencers can be used but these can impart significant back pressure on the process and so waste energy. In addition, reactive silencers cannot follow changes in the frequency of the noise and may also contain chambers in which unwanted material can accumulate, which could affect the performance of the silencer.

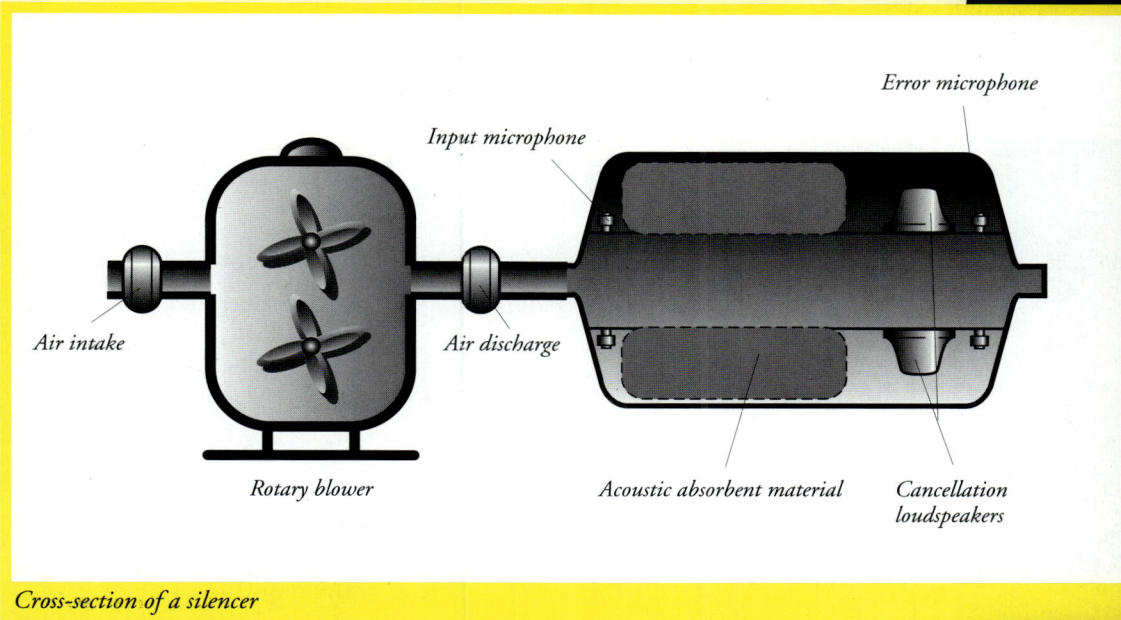

Cross-section of a silencer

The solution

The introduction of a hybrid active silencer (as shown in the illustration) did not impede the air flow and had absorptive packing for the high frequencies and active attenuation for the lower frequencies. The aim of using both methods together is to reduce noise levels from the whole spectrum of frequencies.

The cost

From £3000.

The result

Discharge noise reduced by 42 dB(A). The attenuation - active and absorptive - is shown in the graph.

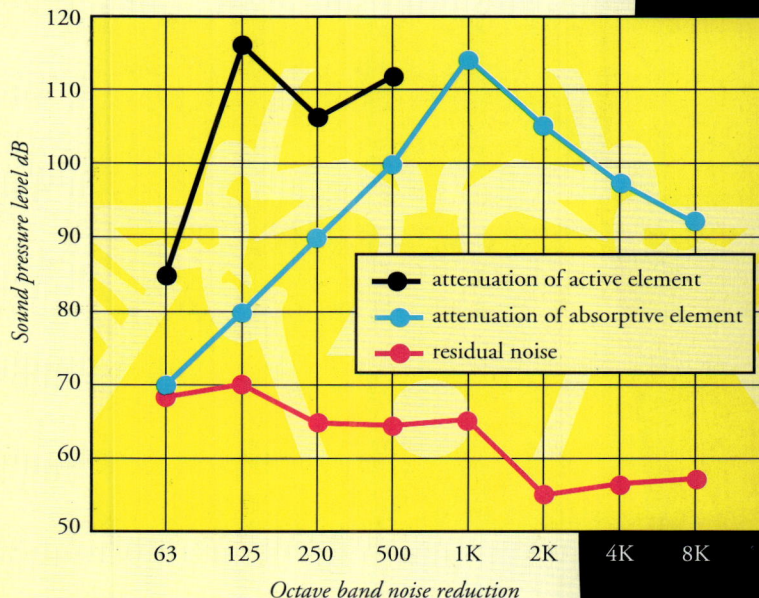

Equipment manufactured by
Digisonix Inc, USA

52
Reducing stone cutting noise

The problem
When using conventional steel saws in stonemasonry, noise levels of over 102 dB(A) can be generated from the impact of each blade segment on the stone and by the mechanical vibrations exciting the blade as it rotates at high speed.

The solution
A number of alternative blades are available. One range includes a laminated blade which absorbs much of the energy by damping the mechanical resonance. The internal construction consists of a layer of copper, sandwiched between two skins of high-tensile steel. The laminated blades are not generally as strong as traditional solid steel blades and are intended for step cutting only.

Another product uses a steel blade which has a series of S-shaped cuts made by laser. These cuts are sealed by cold-rolling the blade. Copper rivets may be added at the base of each gullet. This treatment restricts the propagation of sound waves through the surface of the blade. A laser-cut blade should last as long as a standard blade and is sufficiently robust to be used for full-depth cutting. Generally a 600 mm diameter laminated blade can only be retipped once, whereas a laser-cut blade can be retipped almost as many times as a conventional one.

Maintenance of both types of blade does not require any special skills beyond those needed for traditional blades.

The cost
600 mm diameter: standard blade £55, laminate £115, laser-cut £130.

The result
Measurements taken with these 'quiet blades' during the cutting of similar stone samples show a reduction of between 10 to 12 dB(A) on conventional blade cutting.

Stone cutting

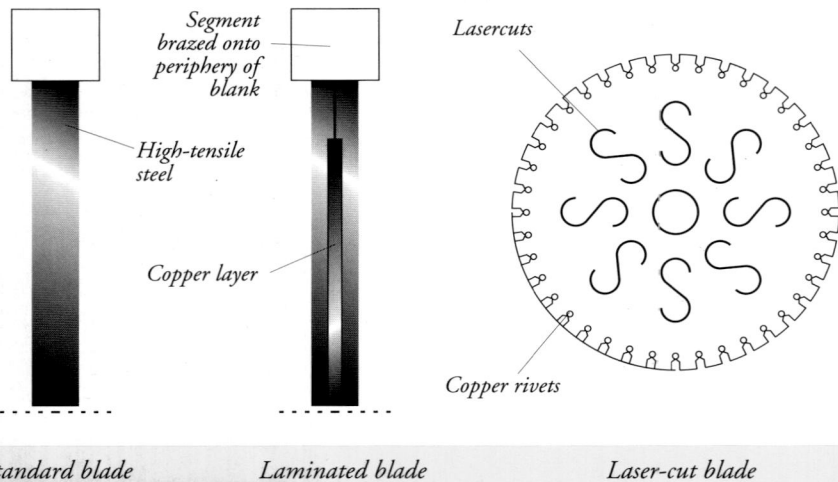

Segment brazed onto periphery of blank

High-tensile steel

Copper layer

Lasercuts

Copper rivets

Standard blade   Laminated blade   Laser-cut blade

Cutting dense sandstone on an ASM profile cutter - straight cut @ 1200 rpm (37 m/s)

Blade type	dB(A)
Standard blade	102.3
Laminated blade	90.6
Laser-cut blade	92.0

Photograph courtesy of
Smiths Limestone Limited

Blades supplied by
Asahi Diamond Industrial UK Company Limited

The problem

Although the process of extruding plastic sections does not normally generate significant noise levels, terminal cut-off saws operating for a few seconds every five minutes can emit very high noise levels.

One company had up to 20 such saws operating in an open production area with employees on the lines being regularly exposed to levels of 100 dB(A) when all the lines were working.

The saws themselves were adequately protected by safety guards. However, as these were all in open steel mesh, they offered no attenuation of sound radiation from the saw blades.

The solution

The sound radiation was effectively impeded by replacing the mesh guards with solid panels, lined on the saw side with acoustically absorptive material. These were fitted both above and below the saw bench. An acoustic strip curtain was hung along the product out-feed, so achieving further attenuation.

The method has since been extended to cut-off saws on new lines, incorporating an acoustically-lined hood to swing into place over a solid base.

The cost

About £650 per hood.

The result

A noise reduction of 15 dB(A) without having either to alter the machine itself or prevent access to the saw.

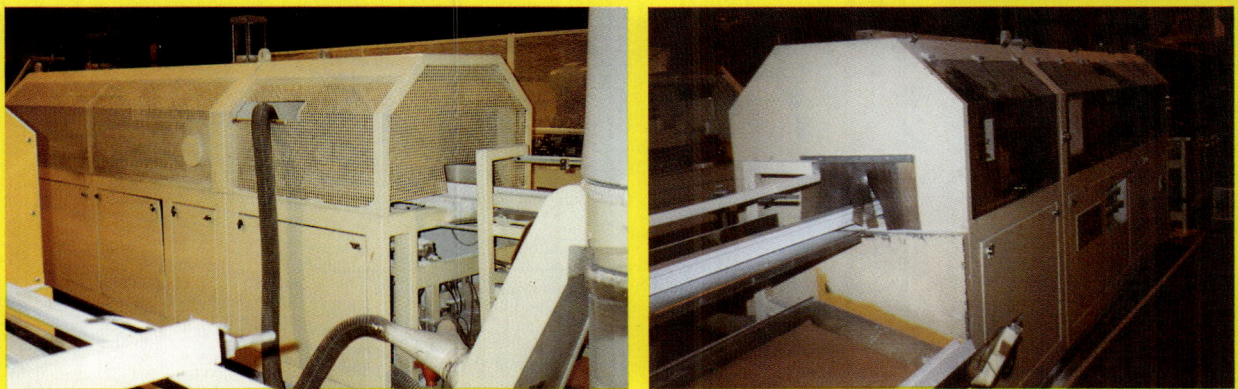

Mesh guards and solid panels

Photographs courtesy of
Caradon Duraplus Limited

The problem

High-pressure steam is essential during the manufacture of Styrene from its constituent hydrocarbons. Unwanted condensation has to be discharged to the outside air through specially-designed steam traps.

This high-temperature, high-pressure steam discharge at high velocities presents a number of hazards to employees working near the traps. Noise levels alone can reach 100 dB(A).

The solution

A major petrochemical company reduced noise levels significantly by fitting diffuser-type silencers to each trap discharge. These silencers operated on the principle that by slowing the jet discharge velocity, the noise output would be reduced. This reduction was achieved by passing the discharge flow into a short pipe of enlarged diameter through a flow-resistive material, typically stainless steel wire wool. The back pressure spread the flow to the larger diameter of the pipe without generating the associated noise-producing turbulence. An added advantage was that by diffusing the discharge it resulted in reduced ground erosion.

The cost

£20.

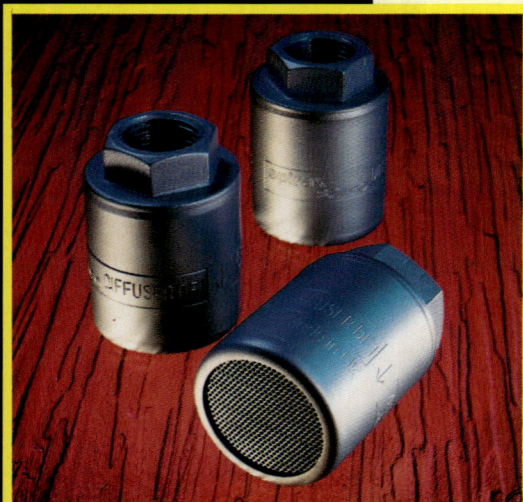

Diffusers

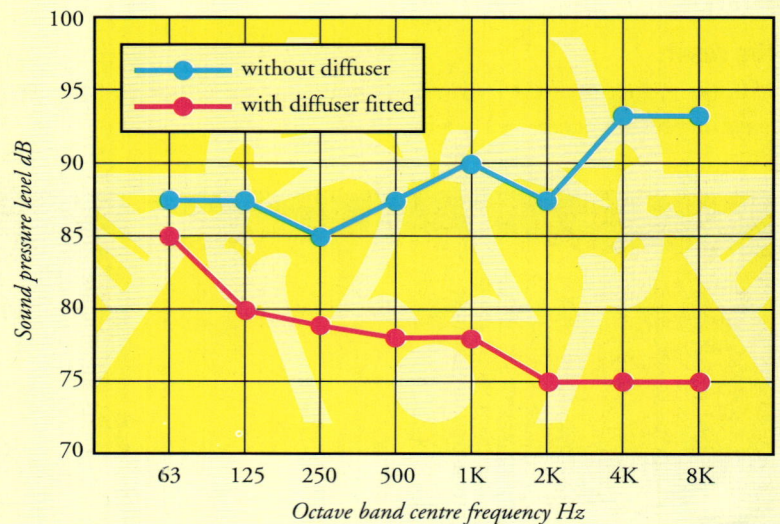

The result

The noise level decreased by 14 dB(A) over a typical discharge cycle, with greater reductions in the mid- to high-frequency bands.

Typical steam discharge with silencer

Photographs courtesy of
BP Chemicals Limited

Diffusers manufactured and supplied by
Spirax-Sarco Limited

The problem

A firm producing fibre mouldings for the automotive industry made use of a number of fibre pulp header tanks, pressurised to ensure a constant flow to the pulp presses. As the level of pulp in a tank was replenished, air pressure rose until a pre-set limit was reached, at which point the pressurised air was released to the atmosphere through a 50 mm port. This discharge contained moisture and, occasionally, some pulp from the header tank.

The tanks were located in working areas of the factory, exposing employees every 10 minutes to noise levels of about 101 dB(A) for about 3 to 4 seconds.

The solution

The solution was to install an absorptive jet discharge silencer, approximately 2 m long, onto the pressure vessel discharge. This had the following features:

- straight-through format so that neither back pressure nor the build up of solid particle carry-over could occur;
- adequate length to absorb the energy;
- a wrapping of thin plastic film to protect the absorbent material from moisture and pulp carry-over;
- removable end connections via a British Standard Pipethread to enable periodic internal cleaning.

Absorptive jet discharge silencer

The cost

£400 per unit.

The result

Details of the measured noise reductions are shown in the table:

Sound pressure level at 1 m from the pressure vessel

Frequency (Hz)	dB(A)	250	500	1000	2000	4000	8000
Before treatment	101	88	92	94	96	94	93
After treatment	91.2	87	88	84	81	85	84
Attenuation (dB)	9.8	1	4	10	15	9	9

Photograph courtesy of
Marley Automotive Components Limited

Equipment designed by
Ian Sharland Limited

56
Modifying acoustically-treated control booths

The problem

The manufacture of steel rod involves a continuous production line to heat square steel billets in a furnace before passing them through a series of about 25 forming stands. They are then coiled and secured as finished wire ready for despatch. In one company, the production line was 200 m long and noise levels averaged about 90 dB(A). The size and nature of the plant rendered acoustic treatment impracticable.

The solution

The company located a series of booths along the line, with each containing the controls necessary for normal operation of that particular section. The booths were either set back from the line or located centrally above it, so that various stages of production could be watched. Where more detailed inspection was required, closed-circuit television was operated from within the booth.

The booths were made from standard acoustic panels with double-glazed plastic windows which reduced the ambient noise levels to about 70 dB(A). The first design was further improved by replacing the original doors with 100 mm thick steel acoustic doors, ensuring good positioning and effective compression of the seals when closed. The air-conditioning units - which emitted noise levels of 80 dB(A) when running on a fast setting - were also removed, with quieter ones being installed in their place.

Views of the control booth

The cost

About £800 per door and £300 for the air-conditioning unit.

The result

A noise reduction of 8 dB(A) from the new doors. Up to 20 dB(A) from the new air-conditioning unit.

Information and photographs courtesy of
Allied Steel Wire

The problem

Timber planer-thicknessing machines usually produce high noise levels, characterised by strong tonal content. The tone was being produced by air turbulence between the cutting head and the table. In one such machine in a joinery shop, it was found that when the machine table was slightly opened away from the cutting head, the noise level appeared to be quieter. It was also discovered that noise from the extraction of dust was capable of making a significant contribution to overall noise levels.

Tests were carried out to determine whether the machine could be set to run continuously at a lower noise level. The findings of those tests are shown in the table below.

Table position	Air supply	Machine condition	Noise level dB(A)
Closed	Full	Cutting	97
		Idling	
25 mm open	Restricted	Cutting	
		Idling	88
Fully open		Idling	90

Views of timber planer-thicknessing machines after adjustment

The solution

The table was set at a 25 mm clearance from the cutter, and airflow was restricted to a volume just sufficient to clear the dust. The productivity, quality and safety of operation were not compromised by increasing the gap between the cutter and the table.

The cost

None.

The result

A noise reduction of between 7 dB(A) with the machine cutting and 13 dB(A) with the machine idling.

Information supplied by
Sound Research Laboratories Limited, Wilmslow

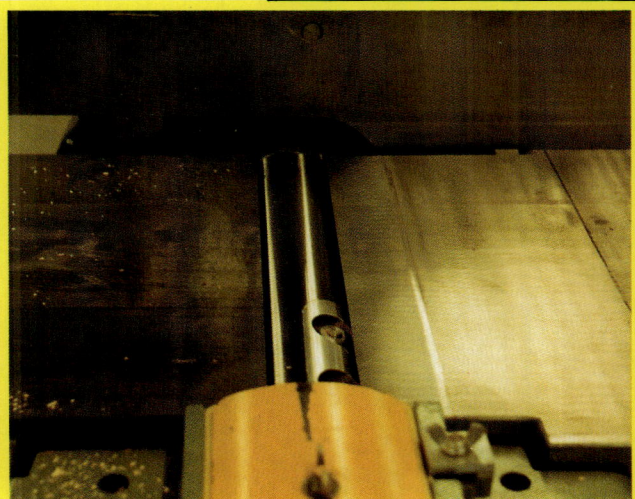

58
Reducing noise in a high-speed transfer press

The problem

The manufacture of components in a multi-stage transfer press begins with the formation of the first stage shape under the first press tool. The tool is then lifted clear of the workpiece. The component is transferred horizontally to the next stage tool where the process is repeated. At each stage the tool ram is pressed down by a cam on the main drive shaft and then lifted by a second cam immediately adjacent on the same shaft (see Figure 1).

In one machine design, the clearance between cam profiles and the up and down ram lifting arms was as much as 100 mm, resulting in a number of metal-on-metal impacts during each rotation of the cam shaft. This produced high noise levels and limited machine efficiency.

The solution

The machine manufacturers replaced the two separate cams with a single cam. This ran against the roller followers fixed on both the lifting arm and pressure actuator plates as indicated in Figure 2. The combination of the new cam profile and the double roller followers enabled operating clearance to be reduced to below 1 mm as well as reduced impact between the cam and rollers. Noise levels were also reduced.

A further modification was to replace the straight spur gears on the main drive train with helical tooth gears.

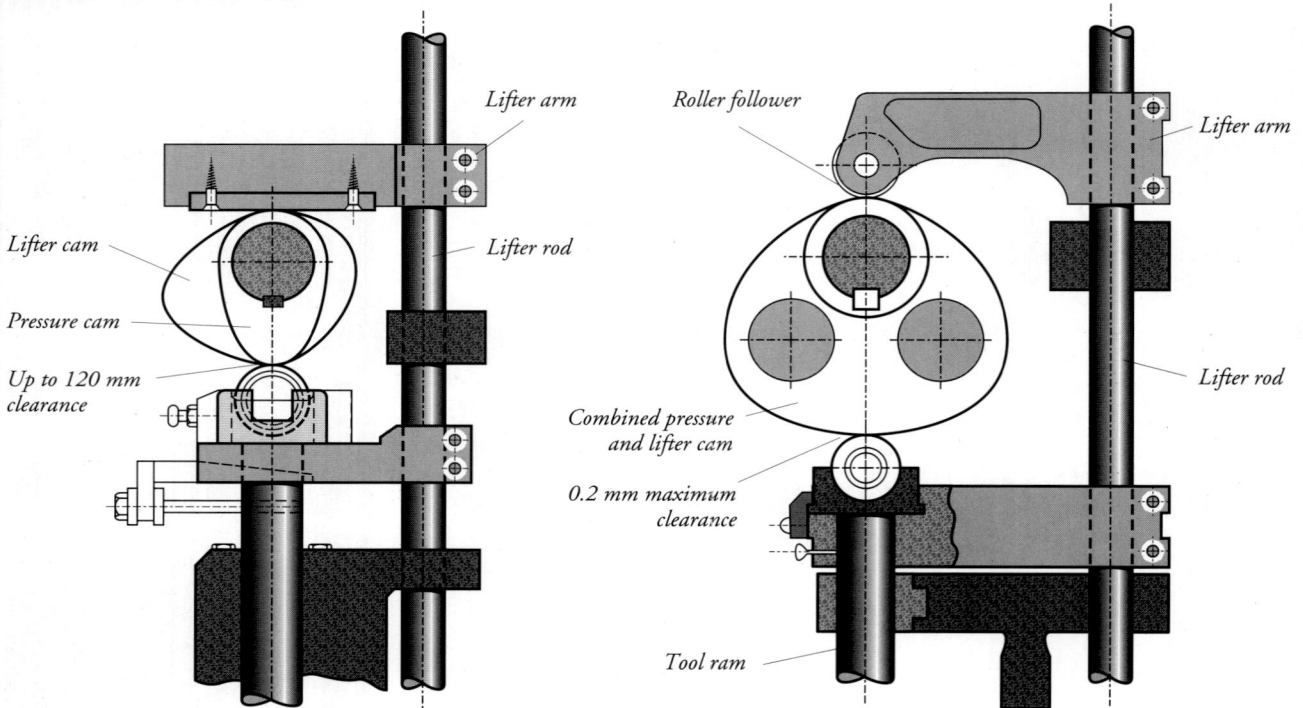

Figure 1 Original two-cam actuator

Figure 2 Modified single-cam actuator

The cost

Costs are not available.

The result

A noise reduction of about 10 dB(A). The press now has higher output.

Information supplied by
Platarg Engineering Limited

The problem

Metal-on-metal impacts are a common and often disproportionately significant source of noise in many metal fabricating shops.

One major motor manufacturer recently installed a cropping machine to automate the finishing of castings, previously carried out by operators using hand-held pneumatic tools.

While the new equipment reduced noise exposure during the process itself, noise was still being generated by the unloading mechanism. A mechanical arm lifted the casting out of the machine, rotated it clear and then dropped it from a height of about 1 m onto a metal chute. The impact of the components onto the chute created peak levels of 100 dB(A).

The solution

A 5 mm layer of wear-resistant rubber matting was applied to the inside surfaces of the chute.

The cost

About £75 per square metre for material, plus installation.

The result

A noise reduction of about 15 dB(A). The treatment also protected the casting and chute from impact damage.

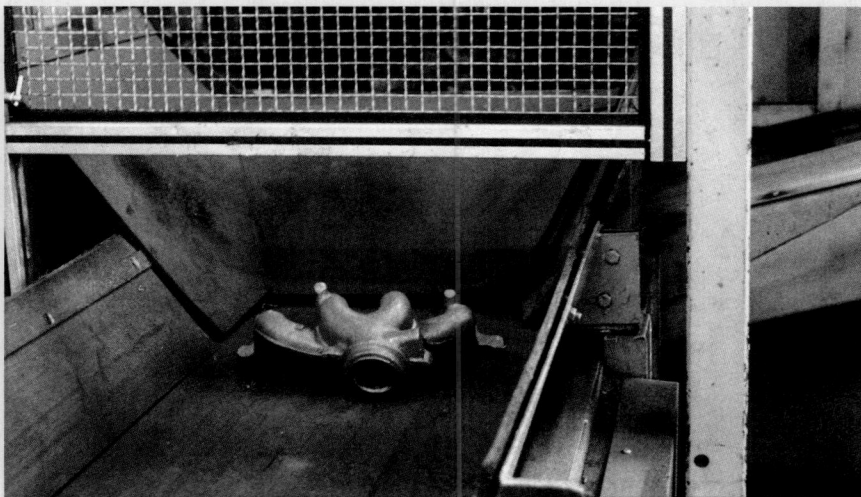

Damped landing chute

Information and photograph supplied by
Ford Motor Company Limited

60
Damping compounds on heavy plates

The problem

A common source of high-level, short-duration noise in the warehousing and distribution industries is caused by hard-wheeled pallet trucks passing over decks of dock levelling plates and scissor lifts during the loading and unloading of road vehicles.

These decks are usually manufactured in a heavy gauge undamped material such as steel or aluminium, with chequered or similar top surfaces for slip resistance. Peak noise levels of up to 108 dB(A) have been measured, arising from the impact between pallet truck wheels and deck surface ridges.

Noise can be reduced by applying damping materials. However the effectiveness of such treatments is limited because:

- the weight and thickness of typical deck plating requires a heavy application of conventional materials;
- low resistance to damage would mean that the damping material has to be applied to the underside of the deck where it is ineffective in reducing the amount of vibrational energy generated in the plate;
- most damping materials have poor weathering properties, limiting their application to decks.

Applying the damping compound

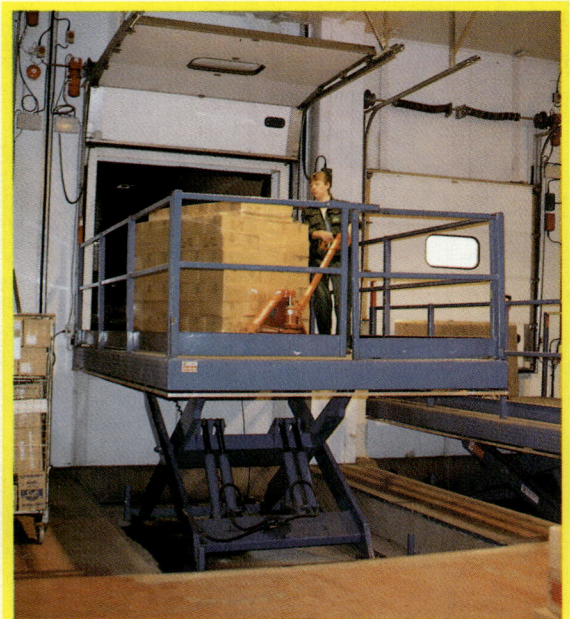

Treated scissor lift

The solution

Recent advances have included the development of blended resin compounds which effectively overcome all of these drawbacks. In particular, one such compound can be applied to the top surface of the deck to improve its anti-slip qualities. It also provides a measure of impact isolation between the truck wheel and the deck.

The method could equally be applied to other instances of noise generated by impact loads on heavy plate structures.

The cost

The approximate cost for the materials to cover 8 square metres of deck is £500.

The result

On controlled tests of the compound it was found that peak noise levels could be reduced to 97 dB(A).

Information and photographs supplied by
Hargreaves Industrial Services Limited

TABLE OF CASE STUDIES BY INDUSTRY

TABLE OF CASE STUDIES BY CONTROL METHOD

Control method	Study
Absorbent lining	21
Absorption	43
Absorption (hung)	46
Absorption (walls)	50
Absorptive/damping/lining	24, 25, 26
Acoustic screen	4, 44, 50
Active noise control	12, 22
Damping	1, 30, 31, 33, 49, 59, 60
Design	9, 36, 42, 57
Design (damping)	29, 52
Design (impact isolation)	7
Design (impact reduction)	6, 58
Design (isolation/duct exhaust)	41
Design (misc)	37, 39
Design (new blower)	3
Design (new engine)	7
Design (new material)	32
Design (nozzle)	2, 3, 16
Design (silencer)	17
Diffuser silencer	54
Enclosure	8, 10, 14, 19, 35, 40, 53
Enclosure (isolation/misc)	48
Enclosure (masonry)	13
Enclosure (strip curtain)	47
Isolation	5, 27, 28, 38, 49
Lagging	20
Maintenance	9, 45
Partition wall	34
Refuge	11, 56
Resonator	23
Silencer	17, 18, 23, 28, 29, 55
Silencer (hybrid active)	51
Silencer (jet)	15

FURTHER READING

General guidance

Noise at work. Noise guide no 1: Legal duties of employers to prevent damage to hearing. Noise guide no 2: Legal duties of designers, manufacturers, importers and suppliers to prevent damage to hearing. The Noise at Work Regulations 1989 HSE Books 1989 ISBN 0 7176 0454 3

Noise at work: Noise assessment, information and control. Noise Guides 3 to 8 HS(G)56 HSE Books 1990 ISBN 0 11 885430 5

HSE free leaflets
(not industry-specific)

Ear protection in noisy firms - Employers' duties explained IND(G)200(L) HSE 1995

Health surveillance in noisy industries - advice for employers IND(G)193(L) HSE 1995

Hear this - a pocket card for workers on ear protection IND(G)201(P) HSE 1995

Introducing the Noise at Work Regulations: a brief guide to the requirements controlling noise at work IND(G)75(L)(Rev) HSE 1989

Listen up! IND(G)122(L) HSE 1992

Noise at work - advice for employees IND(G)99(L)(Rev) HSE 1991

Industry-specific guidance

Priced

Control of noise in quarries HS(G)109 HSE Books 1993 ISBN 0 7176 0648 1

Noise control in the rubber industry HSE Books 1990 ISBN 0 11 885550 6

Noise from pneumatic systems Guidance Note PM 56 HSE Books 1985 ISBN 0 11 883529 7

Noise reduction at buckle folding machines HSE Books 1986 ISBN 0 11 883849 0

Noise reduction at web-fed presses HSE Books 1988 ISBN 0 11 883972 1

Protection of hearing in the paper and board industry HSE Books 1988 ISBN 0 11 883971 3

Free

Noise (Agriculture Safety) AS8(Rev) HSE 1978

Noise enclosure at band re-saws Woodworking Information Sheet No 5 Available free from Woodworking National Interest Group, HSE, 14 Cardiff Road Luton LU1 1PP

Noise from portable breakers IAC/L21 HSC 1986

Noise in construction: further guidance on the Noise at Work Regulations 1989 IND(G)127(L)(Rev) HSE 1993

Noise reduction at band re-saws Woodworking Information Sheet No 4 Available free from Woodworking National Interest Group, HSE, 14 Cardiff Road Luton LU1 1PP

Noise: a ceramics industry booklet IAC/L54 HSC 1992

Protection of hearing in paper and board mills IAC/L46 HSC 1992

Note Many leaflets are available free for single copies, but multiple orders are supplied in priced packs.

GLOSSARY

Absorption   Reduction of sound energy using sound absorbing materials, eg mineral wool.

Acoustic panels   Noise absorbing panels. Usually these are a sandwich construction consisting of dense mineral wool fibre between perforated sheet material on one side (facing the noise source) and plain sheet material on the other.

Active noise control   Where noise is reduced by using an electronic processing system. This system assesses the make-up of the noise and then produces an inverse sound wave, transmitted by a loudspeaker into the path of the noise, so cancelling out the noise.

Actuator   A mechanical device for pushing air in the right direction.

Anodised   Coated with a protective oxide film.

Aspirated venturi   A device to draw in air.

Attenuation   Reduction in sound energy as it travels through a material.

Axial   A fan acting like a propeller in a duct pushing air along.

Billet   A piece of metal.

Decibel (dB) and dB(A)   Unit of sound level using a logarithmic scale. When considering the effect on human hearing, the dB(A) unit is used which takes account of the response of the ear to different frequencies.

Damping   Reduction in vibration (to reduce sound emission).

Destructive interference   See Active noise control.

Frequency (units Hertz, Hz)   The pitch of the sound: the number of oscillations per second of a sound wave.

Fundamental frequency and harmonics   The sound produced by most machines is a complex mixture of frequencies. Some consist of a fundamental frequency (the lowest frequency of the mixture) and a series of overtones, or harmonics, with frequencies that are simple multiples of the fundamental.

Helical   Shaped like a cylindrical spiral.

Insulation   A material which reduces sound passing through its thickness.

Noise exposure   A measure of the total sound energy a person is exposed to. It is dependent on both the sound pressure level to which the person is exposed and the time over which the exposure occurs.

Octave band   The audible frequency range from 20 to 20 000 Hz is divided into 10 octave bands. These octave bands are arranged so that the centre of one band is twice the frequency of the previous band (one octave higher), ie 250 Hz, 500 Hz, 1000 Hz, 2000 Hz etc.

Pure tone   A single frequency noise.

Radial   A fan pulling air into the centre of a vortex. The action of the blades throws the air to the sides of the duct.

Ram   Part of a piston.

Resonance   The frequency at which a machine or any part of it naturally vibrates.

Reverberation   The sound from a noise source being reflected from walls, floors, ceilings and obstacles within a room before being absorbed. If many reflections occur before absorption, the reflected noise adds to that directly from source noise, increasing the overall noise level.

Silencer (absorptive and reactive)   A device which reduces the noise travelling along a duct or pipe either using sound absorbing materials (absorptive silencer) or using chambers of differing cross-sections and baffles (reactive silencer), eg a car exhaust silencer.

Sound intensity   A measure of sound energy flow.

Sound intensity survey   A series of measurements of sound intensity designed to highlight where sound is being emitted from a machine.

Sound power level   A measure of the total acoustic power produced by a noise source.

Sound pressure level   A measure of the magnitude of sound pressure fluctuations at a specific location.

Transmission loss   The extent to which sound energy will be reduced when transmitted through an insulating material.

Tonal   Sound that has a plainly audible single frequency components, such as a whistle, whine or hum.

Wavelength   The distance between two identical points on a pure-tone sound wave, measured along the direction of propagation of the wave, eg the distance from crest to crest.

ACKNOWLEDGEMENTS

The Health and Safety Executive acknowledges the generosity of the following companies who supplied information for case studies or photographs/illustrations to help compile this publication:

Air Control Installations (Chard) Limited, Chard
Ajax Health and Safety Services, Redditch
Allied Steel Wire, St Mellons, Cardiff
Alumasc-Grundy Limited, Burton-upon-Trent
Asahi Diamond Industrial UK Company Limited, Crawley
Ash Peace Limited, Martock
Atlas Copco Tools Limited, Hemel Hempstead
Bass Mitchells and Butlers, Cape Hill Brewery, Birmingham
BP Chemicals Limited, Port Talbot
Brico Engineering Limited, Coventry
British United Shoe Machinery Limited, Leicester
Caradon Duraplus Limited, Toddington, Cheltenham
Clarks International, Street
Coventry City Council (Environmental Services Department)
Digisonix Inc., Wisconsin, USA (London office - London Road, SE1)
Dunham-Bush Limited, Havant
EFCO Limited, Woking
EMD Services, Bourne
Euro Acoustics Limited, Barrow-upon-Soar, Loughborough
Ford Motor Company Limited, Leamington Spa
Fort Vale Engineering Limited, Parkfield Works, Nelson
Glynwed Foundries, Ketley, Telford
Hallam Polymer Engineering Limited, Dronfield
Hargreaves Industrial Services Limited, Woodlesford, Leeds
Ian Sharland Limited, Winchester
ICI Chemicals Plc, Hallam, Bristol
Industrial Noise and Vibration Centre, Slough
Industrial Noise Services Limited, Stourbridge
Institute of Sound and Vibration Research, Southampton
ISVR Consultancy Services, University of Southampton
John Brown Plastics Machinery Ltd, (Cumberland Engineering Division), Stroud
John Crane UK Limited, Slough
Jones Engineering, Westwoodside, Doncaster
Joseph Rhodes Limited, Wakefield
Lucas Industries Noise Centre, Shirley, Solihull
Lyons Seafoods Limited, Warminster
Mainline Industrial Products Limited, Romsey
Marley Automotive Components Limited, Bristol
Mato Industries Limited, Flint
Medica Packaging Limited, Crewe
Meech Exair, Witney
Micronizing Company UK Limited, Framlingham
The Noise Control Centre, Melton Mowbray
Noise Reduction Limited, Eastleigh

Platarg Engineering Limited, London W7
Preedy Glass Limited, London NW10
R S Allsopp and Associates, Melton Mowbray
Redland Precast, Barrow-upon-Soar, Loughborough
SmithKline Beecham Consumer Healthcare, Brentford
Smiths Limestone Limited, Broadway
Sound Research Laboratories Limited, Wilmslow
Sound Research Laboratories Limited, Little Waldingfield, Sudbury
Spirax-Sarco Limited, Cheltenham
Thames Water Utilities Limited, Reading
Trellex Limited, Rugby
Vernon Cole Associates, Little Bookham

Printed and published by the Health and Safety Executive   C100   9/95